Potassic Igneous Rocks
and Associated Gold-Copper Mineralization

Springer
Berlin
Heidelberg
New York
Barcelona
Hong Kong
London
Milan
Paris
Singapore
Tokyo

Daniel Müller · David I. Groves

Potassic Igneous Rocks and Associated Gold-Copper Mineralization

3rd, Updated and Enlarged Edition

With 61 Figures and 68 Tables

Springer

Dr. Daniel Müller
Institute for Mineralogy
Mining-Academy Freiberg
Brennhausgasse 14
09596 Freiberg, Germany

Prof. Dr. David I. Groves
The University of Western Australia
Centre for Strategic Mineral Deposits
Department of Geology and Geophysics
Nedlands, W.A. 6907, Australia

The cover shows an overview of the Bajo de la Alumbrera porphyry copper-gold deposit before the commencement of mining operations.

The first two editions have been published in the series Lecture Notes in Earth Sciences.

ISBN 3-540-66371-1 3rd, updated and enlarged ed. Springer-Verlag Berlin Heidelberg New York
ISBN 3-540-62075-3 2nd, updated and enlarged ed. Springer-Verlag Berlin Heidelberg New York

Library of Congress Cataloging-in-Publication Data
Müller, Daniel, 1962 – Potassic igneous rocks and associated gold-copper mineralization / Daniel Müller, David I. Groves. – 3rd updated and enl. ed. p. cm. Includes bibliographical references and index. ISBN 3-540-66371-1 (hardcover) 1. Rocks, Igneous. 2. Gold ores. 3. Copper ores. I. Title. II. Groves, D.I. QE461.M773 1999 552'.1 21–dc21, 99-040223

© Springer-Verlag Berlin · Heidelberg 1995, 1997, 2000
Printed in Germany

The use of general descriptive names, registered names, trademarks, etc. in this publication does not imply, even in the absence of a specific statement, that such names are exempt from the relevant protective laws and regulations and therefore free for general use.

Cover: Erich Kirchner, Heidelberg
Typesetting: Camera-ready by the authors
SPIN 10707701   32/3136-5 4 3 2 1 0 – Printed on acid-free paper

Preface

In recent years, there has been increasing interest from geoscientists in potassic igneous rocks. Academic geoscientists have been interested in their petrogenesis and their potential value in defining the tectonic setting of the terranes into which they were intruded, and exploration geoscientists have become increasingly interested in the association of these rocks with major epithermal gold and porphyry gold-copper deposits. Despite this current interest, there is no comprehensive textbook that deals with these aspects of potassic igneous rocks.

This book redresses this situation by elucidating the characteristic features of potassic (high-K) igneous rocks, erecting a hierarchical scheme that allows interpretation of their tectonic setting using whole-rock geochemistry, and investigating their associations with a variety of gold and copper-gold deposits, worldwide. About half of the book is based on a PhD thesis by Dr Daniel Müller which was produced at the Centre for Strategic Mineral Deposits (former ARC Key Centre) within the Department of Geology and Geophysics at The University of Western Australia under the supervision of Professor David Groves, the late Dr Nick Rock, Professor Eugen Stumpfl, Dr Wayne Taylor, and Dr Brendan Griffin. The remainder of the book was compiled from the literature using the collective experience of the two authors. The book is dedicated to the memory of Dr Rock who initiated the research project but died before its completion.

Sincere thanks are due to our colleagues and friends at the former Key Centre for providing a stimulating environment in which to do the research and to write the first drafts of the book. Gratitude is also expressed to former colleagues at Placer International Exploration Inc. in Santiago, PT North Mining Indonesia in Jakarta, and North Limited in Parkes and Kalgoorlie for encouraging the completion and revision, respectively, of the book. Professor Eugen Stumpfl is also sincerely thanked for his hospitality and assistance in the early stages of the research recorded in the book. We are particularly indebted to Dr Susan Ho who assisted in the editing of the book, produced the camera-ready copy, and organised the subject index. Without her help, the book would not have been completed on schedule, if at all. Col Steel is also thanked for his excellent drafting of the more complex maps displayed in the book. Dr Wolfgang Engel of Springer-Verlag is also thanked for his encouragement of the project.

The following colleagues also provided support, contributed ideas, shared authorship on papers, and/or provided unpublished information:
Anita-Kim Appleby, Phil Blevin, Eric Bloem, David Bowes, Mari Carrizo, Andi Zhang, Megan Clark, Hilko Dalstra, Alan Edgar, Michael Farrand, Peter Forrestal, Richard Förster, Michael Gareau, Musie Gebre-Mariam, Rich Goldfarb, Sue Golding, Eliseo González-Urién, Roland Gorbatschev, Lalu Gwalani, Greg Hall, Adolf Helke, Paul Heithersay, Bruce Hooper, Abraham Janse, Rod Jones, Roger Jones, Imants Kavalieris, Jeffrey Keith, Megan Kenny, David Keough, Rob Kerrich, Ken Lawrie, Bob Love, Neal McNaughton, Ian Miles, Claudio Milliotti, Aberra Mogessie, Brian Morris, Gregg Morrison, Peter Neumayr, Juhani Ojala, Julian Pearce, Joe Piekenbrock, David Quick, David Radclyffe, Rob Ramsay, Hector Salgado, Steve Sheppard, Richard Sillitoe, Henning Sørensen, Hernan Soza, Jon Standing, Joe Stolz, Shen-Su Sun, William Threlkeld, Spencer Titley, Linda Tompkins, Takeshi Uemoto, Ignacio Ugalde, Theo van Leeuwen, Gianpiero Venturelli, Marcial Vergara, Richard Vielreicher, Mike Wheatley, Noel White, Rohan Wolfe, Doone Wyborn, and Derek Wyman.
Elsevier permitted the use of Figures 5.4, 5.5, 5.6, 6.2, 6.3, 6.4, 6.5, 6.6 and 6.16, and Table 6.1.

Daniel Müller, Freiberg
David I. Groves, Perth

Table of Contents

Abbreviations

Short definitions of abbreviations that are used frequently in this book.

HFSE   High-field-strength elements (e.g. Ti, Y, Zr, Nb, Hf, Ta). These elements are characterised by small atomic *radii* and high atomic charges. They are normally accommodated into the lattice sites of titanites and apatites. Subduction-derived potassic igneous rocks have very low abundances of high-field strength elements, while those generated in within-plate tectonic settings have high concentrations. Generally, high abundances of high-field strength elements are considered to reflect deep asthenospheric magma sources.

LILE   Large-ion lithophile elements (e.g. K, Rb, Sr, Cs, Ba). These elements are characterised by large ionic *radii* and low atomic charges. They are not readily accommodated into the lattice of upper-mantle minerals and are mantle-incompatible. Large-ion lithophile elements are strongly partitioned into the first melt increments during small degrees of partial melting. They are commonly sited in hydrous minerals such as biotites, phlogopites and amphiboles. Potassic igneous rocks are enriched in large-ion lithophile elements.

LOI   Loss on ignition. This is the proportion of mass lost (as volatiles) when rock powder is heated at about 1100°C in a furnace for an hour or more. It usually corresponds to the total content of H_2O, CO_2, and S.

LREE   Light rare-earth elements (e.g. La, Ce, Nd). They are part of the lanthanides (atomic numbers 57-71), commonly equated by petrologists with the rare-earth elements. The light rare-earth elements represent those lanthanides with the lower atomic numbers and the larger atomic *radii* due to the "lathanide-contraction" with increasing atomic numbers. They are mantle-incompatible and are preferentially enriched in the first melt increments during low degrees of partial melting. Light rare-earth elements abundances tend to increase during the process of differentiation. They are normally accommodated in the lattice sites of clinopyroxenes and apatites. Potassic igneous rocks are enriched in light rare-earth elements.

mg#   Molecular $Mg/(Mg+Fe^2)$. Unless otherwise indicated, this value is calculated in this book with molecular $Fe^2/(Fe^2+Fe^3)$ set at 0.15, a common ratio in potassic igneous rocks.

MORB   Mid-ocean ridge basalt. These basalts occur at the spreading centres of the mid-ocean ridges. They are derived from partial melting of a depleted mantle source, and their geochemical composition is tholeiitic with low concentrations in mantle-incompatible trace elements.

OIB   Oceanic island basalt. These basalts are generally regarded as being derived from chemically-anomalous mantle sources and to represent hotspot magmatism where asthenospheric mantle plumes impinge on the surface of oceanic crust. Their geochemistry is alkaline with characteristic enrichments in mantle-incompatible elements such as potassium. Examples where oceanic island basalts occur are Tristan da Cunha and Gough Island in the South Atlantic.

PGE   Platinum-group elements (e.g. Pt, Pd). The group of precious metallic elements comprising ruthenium, rhodium, palladium, osmium, iridium, and platinum.

Notes regarding tables of geochemical data

Major elements are listed in order of decreasing valency from SiO_2 to K_2O, followed by P_2O_5, LOI, SrO, BaO, Cl, and F where data are available and the abundance is sufficient for the element to be considered a major, rather than trace, element.

Trace elements are listed in order of increasing atomic number.

Where data are available, *precious metals* have been separated and listed in order of increasing atomic number.

1 Introduction

1.1 Preamble: Potassic Igneous Rocks and Their Importance

Potassic igneous rocks occur in many different tectonic settings (e.g. Rock 1991; Foley and Peccerillo 1992), and include a variety of compositions ranging from shoshonites associated with calc-alkaline volcanic rocks to ultrapotassic leucitites (Foley and Peccerillo 1992; Peccerillo 1992). They are of increasing economic interest due to their association with mineralization, and are of tectonic significance because of their potential value in reconstructing the tectonic setting of ancient terranes.

On the *economic* front, potassic igneous rocks are now established as being closely related to certain types of gold and base metal deposits (Mitchell and Garson 1981; Heithersay et al. 1990; Kavalieris and Gonzalez 1990; Richards 1990a; Setterfield 1991; Mutschler and Mooney 1993). Some may even be intrinsically enriched in Au and platinum-group elements (PGE) (Wyborn 1988; Müller et al. 1992a, 1993a). Some of the world's largest volcanic- and intrusion-hosted gold and copper-gold deposits are intimately related to potassic igneous rocks. For example, the world-class epithermal gold deposits at Ladolam and Porgera, Papua New Guinea (Moyle et al. 1990; Richards 1990a; White et al. 1995), El Indio, Chile (Jannas et al. 1990), Baguio, Philippines (Cooke et al. 1996), Cripple Creek, USA (Thompson 1992), and Emperor, Fiji (Anderson and Eaton 1990; Setterfield 1991) are all hosted in potassic calc-alkaline or alkaline rocks. Similarly, there are equivalent host plutons associated with the porphyry gold-copper deposits at Bingham, USA (Keith et al. 1997), Bajo de la Alumbrera, Argentina (Guilbert 1995; Müller and Forrestal 1998), Cadia and Goonumbla, NSW, Australia (Heithersay et al. 1990; Müller et al. 1994), Ok Tedi, Papua New Guinea (Rush and Seegers 1990), and Grasberg, Indonesia (Hickson 1991). Although emphasized in previous editions of this book, this association has not been stressed in previous reviews on porphyry and epithermal styles of mineralization. However, Sillitoe (1997), in a recent review of world-class gold-rich porphyry and epithermal deposit in the circum-Pacific region, emphasizes the association of many of these giant deposits with potassic igneous rocks: their location and approximate gold content is shown in Figure 1.1. Sillitoe (1997) points out that,

even excluding those deposits associated with potassic igneous rocks of other magmatic associations (e.g. high-K calc-alkaline suites), about 20% of the large gold
deposits are associated with shoshonitic and alkaline rocks, which are unlikely to
exceed 3% by volume of circum-Pacific igneous rocks. Importantly, Sillitoe (1997)
listed this association as one of only four criteria favourable for *both* world-class
porphyry and epithermal gold-rich deposits in the circum-Pacific region, and at-
tributes this association to the partial melting of stalled lithospheric slabs in the mantle, immediately following collision or arc migration, as a preferred mechanism to
promote oxidation of mantle sulphides and the release of gold.

   In addition, less important examples from an economic viewpoint include the
Lamaque stockwork at Val d'Or, Quebec, Canada (Burrows and Spooner 1991), the
porphyry copper-gold mineralization associated with shoshonitic rocks in British
Columbia, Canada (Barrie 1993; Kirkham and Margolis 1995; Kirkham and Sinclair
1996), the Jinjushan epithermal gold deposit, Lower Yangtze region, China (Zhou et
al. 1996), and Prospector Mountain, Yukon, Canada (Glasmacher and Günther 1991).
Subduction-related cobalt-nickel mineralization in northeast Scotland, formed during an arc-continent collision during the Grampian Orogeny, is also associated with
potassic igneous rocks (Dunham 1974; Mitchell and McKerrow 1975). Potassic igneous rocks are thus becoming important exploration targets in their own right.

   On the *tectonic* front, potassic igneous rocks have been recognized as an important and integral component of destructive continental margins (e.g. Hatherton and
Dickinson 1969; Morrison 1980; Saunders et al. 1980; Carr 1998). Although there

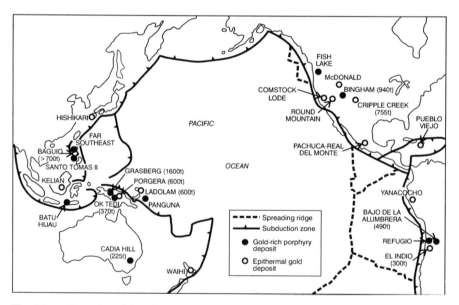

Fig. 1.1.   Location of the largest gold-rich porphyry and epithermal gold deposits of the
circum-Pacific region. Deposits with approximate tonnes of contained gold given in brackets
are those associated with high-K igneous rocks. Adapted from Sillitoe (1997).

are exceptions (Arculus and Johnson 1978), arc-related potassic igneous rocks are generally younger, stratigraphically higher, and erupted further from the suture than less potassic rocks, implying that they form at greater depth in a Benioff Zone. This has led to the use of potassic igneous rocks to attribute arc-like tectonic affinities to ancient terranes (Brooks et al. 1982; Barley et al. 1989; Wyman and Kerrich 1989a, 1989b; Wyborn 1992).

It is becoming important, whether in improving exploration models for ancient mineral deposits, or in reconstructing ancient terranes, to be able to distinguish the tectonic settings in which ancient potassic igneous rocks were generated.

1.2 Scope of Book

As part of the "*Drosophila* of igneous petrology" (Barker 1983, p. 297), potassic igneous rocks have gained much attention among petrologists worldwide, mainly due to their distinct geochemistry, and many geoscientists still consider them as petrological curiosities with an obscure petrogenesis. In the past, a plethora of genetic hypotheses and a large number of local names for potassic igneous rocks from different localities have been created (see reviews by Sørensen 1974; Peccerillo 1992). This has produced some confusion in the literature.

This book reviews the geochemical and petrological characteristics of the potassic igneous rock clan, and it investigates the different tectonic settings in which these rocks occur. The authors seek to provide an overview and a classification of those rocks, and to elucidate the geochemical differences between barren and mineralized potassic igneous complexes. Many epithermal gold and porphyry copper-gold deposits are hosted by high-K rocks. Therefore, this book is not only relevant to the academic petrologist working on alkaline rocks, but also to the exploration geologist prospecting for epithermal gold and/or porphyry copper-gold deposits in modern and ancient terranes.

2   Definitions and Nomenclature

2.1.   Historical Perspective of Potassic Igneous Rocks

Potassic igneous rocks were originally recognized in the late 19th century by Iddings (1895), who described some orthoclase-bearing basalts from the Yellowstone Park, Wyoming, and coined the term "shoshonite". In the last century, petrologists generated many names for potassic igneous rocks which were either based on their mineralogy or, more commonly, based on the locality of their occurrence. The practice was to name a new rock after a place where it occurred — the type locality. These different names for essentially similar rocks from different localities led to great confusion (Sørensen 1974; De Wit 1989; Rock 1991; Peccerillo 1992).

The first attempts to explain the petrogenesis of potassic magmatism date back to the beginning of the 20th century when Daly (1910) explained potassic melts as products of the assimilation of carbonate sediments by uprising basaltic magmas. Rittmann (1933) adopted this hypothesis in order to explain the potassic magmatism of the Vesuvius volcano and the *Mediterranean Series*, as potassic igneous rocks were named at that time (Peccerillo 1992), with the assimilation of carbonates by evolved trachytic magmas. This model was widely accepted until the 1960s, although it was unable to explain the potassic magmatism in the East African Rift, where carbonates are absent. However, Savelli (1967) was able to demonstrate that potassic magmas have much higher abundances of large-ion lithophile elements (LILE) and mantle-compatible elements, such as Cr and Ni, than do both carbonates and basalts. Therefore, the *assimilation model* appeared rather unlikely and alternative explanations were developed. One of these was the *zone-refining model* proposed by Harris (1957). This model was adapted from the steel industry, where the process of zone-refining was used to purify metal bars. Harris (1957) suggested that a mantle plume would rise adiabatically by melting the roof rocks at its top and by crystallizing minerals at its base. This process would allow the rising melt to incorporate all the mantle-incompatible impurities such as LILE and light rare-earth elements (LREE). As a result, the migrating melt would become progressively enriched in these elements and gain a potassic composition. Another model to explain potassic magmatism was based on observations from trace-element modelling (Kay and Gast

1973), which implied that the enrichments in LILE and LREE in potassic igneous rocks were an effect of very low degrees of partial melting (i.e. melt increments of < 1 vol. %) of a garnet-peridotite in the upper mantle.

However, the advent of the *concept of mantle metasomatism* (e.g. Menzies and Hawkesworth 1987) represented a major breakthrough in understanding of the petrogenesis of potassic igneous rocks (Peccerillo 1992). Direct evidence for heterogeneous mantle compositions on a small scale was provided by the petrographic studies of mantle xenoliths from deep-seated kimberlite and alkali-basalt eruptions (e.g. Harte and Hawkesworth 1989) which revealed the presence of LILE-bearing minerals such as phlogopite and apatite within the peridotites of the upper mantle. These minerals, which may occur either in veins or dispersed within the mantle peridotite (Bailey 1982), are believed to have been metasomatically introduced by volatile- and LILE-enriched *fluids* and/or LILE- and LREE-enriched alkalic *melts* (see Chap. 3). The nature and origin of these metasomatizing agents are still under debate (Peccerillo 1992).

Interest in the petrogenesis of potassic magmas has, for many years, been aimed at describing specific occurrences and explaining the differences between these and normal basalts (Foley and Peccerillo 1992). Potassic igneous rocks have features in common with both the alkaline and calc-alkaline rock associations, but also have geochemical characteristics that distinguish them from the other rock associations and, therefore, they must be considered a distinct rock association (Morrison 1980).

In the first comprehensive study of potassic igneous rocks from different localities, undertaken by Sahama (1974), only ultrapotassic rocks were considered and these were divided into kamafugitic and orenditic types. However, the peralkaline orenditic ultrapotassic igneous rocks and the kamafugites, which are represented by groups I and II in Foley et al. (1987), are not considered further in this study (see definitions in Sects 2.5.2, 2.5.3). Modern studies of shoshonites and potassic igneous rocks (e.g. Morrison 1980) re-established their importance as a distinctive group among the spectrum of igneous rocks. High-K rocks such as shoshonites have been formally incorporated in numerous classification schemes (Peccerillo and Taylor 1976a), including that recommended by the IUGS Subcommission (Le Maitre et al. 1989).

2.2   Potassic Igneous Rocks as an Umbrella Term

The *potassic igneous rocks,* as considered here, comprise volcanic, hypabyssal and plutonic rocks. Petrographically, potassic igneous rocks range from basalts and andesites to trachytes, which normally have porphyritic textures with phenocrysts of leucite, plagioclase, alkali feldspar, clinopyroxene, olivine, phlogopite, and/or amphiboles. The term "potassic igneous rocks" is used in this book as an umbrella term to describe those rocks which are more K-rich than typical igneous rocks (i.e. K > Na). The term includes subduction-related high-K calc-alkaline rocks and shoshonites, high-K rocks from within-plate tectonic settings, hypabyssal high-K rocks such as shoshonitic and alkaline lamprophyres (cf. Rock 1991), and the orogenic ultrapotassic rocks (group III of Foley et al. 1987), which are defined in Chapter 2.5.

2.3 Shoshonites

Shoshonites (sensu stricto) are potassic igneous rocks which occur in subduction-related tectonic settings (Morrison 1980). They are commonly formed during the late stage of arc-evolution, being erupted after the low-K tholeiites and calc-alkaline rock series. Although there are a few exceptions, they are commonly most distant from the trench and are erupted above the deepest parts of the Benioff Zone. The shoshonite association is geochemically defined by high total alkalies ($Na_2O + K_2O$ > 5 wt %), high K_2O/Na_2O ratios (> 0.6 at 50 wt % SiO_2, > 1.0 at 55 wt % SiO_2), low TiO_2 (<1.3 wt %), high but variable Al_2O_3 (14–19 wt %), and a strong enrichment in LILE and LREE (Morrison 1980). Basalts and basaltic andesites predominate in the shoshonite association. Shoshonites have porphyritic textures with phenocrysts of plagioclase, clinopyroxene, olivine, phlogopite and/or amphiboles in a very fine-grained, commonly glassy, groundmass consisting mainly of alkali feldspar (sanidine), plagioclase, and clinopyroxene (Morrison 1980).

2.4 Shoshonitic and Alkaline Lamprophyres

Lamprophyres (Greek *lampros, porphyros*: glistening porphyry) form an extremely heterogeneous group of predominantly hypabyssal alkaline igneous rocks which occur in a wide variety of geological settings throughout the world (Rock 1991). In many localities, lamprophyres are associated with granitic, shoshonitic, syenitic, or carbonatitic magmatism (Rock 1991). Several contradictory classifications for lamprophyres have been used over the past century. However, lamprophyres have been comprehensively defined by Rock (1987, 1991) as hypabyssal, melanocratic igneous rocks with porphyritic textures carrying only mafic phenocrysts, essentially phlogopite-biotite and/or amphibole with minor olivine. Phlogopite or biotite phenocrysts are commonly zoned, with dark brown Fe-rich rims and pale yellow Mg-rich cores (Rock et al. 1988b). Felsic minerals are generally restricted to the groundmass. However, quartz xenocrysts are common due to the volatile-driven rapid uprise of lamprophyric magmas (Rock 1991). Lamprophyres are also characterized by battlemented phlogopites and globular structures, which are due to the segregation of late-stage melts — commonly with evolved syenitic compositions — into vugs within the crystal mush (Foley 1984; Rock 1991). The rocks occur as dykes, sills, plugs, stocks, or vents and associated intrusive or explosion breccias.

Geochemically, lamprophyric magmas have primitive compositions, as shown by high mg# [where mg# = molecular $Mg/(Mg+Fe^2)$, with molecular $Fe^2/(Fe^2+Fe^3)$ set at 0.15, a common ratio in potassic igneous rocks] and high Cr, Ni, and V contents. They are typically enriched in LILE, LREE, and volatiles such as CO_2, H_2O, F, and Cl (Rock 1987; Rock et al. 1990), which are sited in the lattice of hydrous minerals such as amphiboles or micas, or hosted by primary carbonates, zeolites, epidotes, fluorites, or sulphates (Rock et al. 1988b).

The lamprophyre clan comprises shoshonitic (calc-alkaline), alkaline, and ultramafic lamprophyres, as well as lamproites and kimberlites (Rock 1991). Only the first two varieties, shoshonitic and alkaline lamprophyres with high K_2O (> 1 wt %) and SiO_2 contents (> 40 wt %), are considered in this study. *Shoshonitic lamprophyres* with groundmass plagioclase > alkali-feldspar are further divided into the amphibole-bearing spessartites and mica-bearing kersantites, whereas those with alkali-feldspar > plagioclase are divided into the amphibole-bearing vogesites and mica-bearing minettes (Rock 1977). *Alkaline lamprophyres* are normally characterized by biotite or phlogopite phenocrysts in a groundmass with alkali-feldspar > plagioclase (e.g. Müller et al. 1992a, 1993a). Many alkaline lamprophyres would be classified as volatile-rich alkali basalts or basanites when plotted on the Na_2O+K_2O versus SiO_2 diagram (Rock 1991) which is recommended by the IUGS Subcommission on Igneous Rocks Systematics (cf. Le Maitre 1989).

2.5   Ultrapotassic Rocks

2.5.1   Introduction

Ultrapotassic rocks are defined by using the chemical screens K_2O > 3 wt %, MgO > 3 wt %, and K_2O/Na_2O > 2 for whole-rock analyses (Foley et al. 1987). They can be further divided into four groups:

- Group I (e.g. the Gaussberg lamproites, Antarctica), characterized by low CaO (< 8 wt %), Al_2O_3 < 12 wt %, Na_2O < 2 wt %, and high mg# (~ 60–85).
- Group II (e.g. the kamafugites of the Toro-Ankole region, East African Rift), characterized by very low SiO_2 (< 40 wt %) and high CaO (> 10 wt %).
- Group III (e.g. the orogenic ultrapotassic rocks of the Roman Province), which occur in orogenic areas and have high CaO (> 5 wt %), high Al_2O_3 (> 12 wt %), and low mg# (~ 40–65).
- Group IV showing transitional chemical characteristics between groups I and III (Foley et al. 1987).

2.5.2 Lamproites

Lamproites commonly occur as volumetrically small vents, pipes, or dykes, and form a group within the potassic igneous rock clan which shares certain petrogenetic aspects with the alkali basalts, kimberlites, and lamprophyres (Bergman 1987). Lamproites have achieved increased economic importance since the discovery of the diamond-bearing Argyle lamproite pipe, Western Australia (Rock 1991). Their occurrence is restricted to within-plate settings (Mitchell 1986; Mitchell and Bergman 1991).

Lamproites are derived by small degrees of partial melting of a phlogopite-harzburgite mantle source under reducing conditions, and two varieties, olivine lamproite and leucite lamproite, may be distinguished (Edgar and Mitchell 1997). Lamproites are normally characterized (Prider 1960; Mitchell and Bergman 1991; Peccerillo 1992) by the presence of rare minerals such as titanian phlogopite, potassic richterite, leucite, jeppeite, sanidine, aluminium-poor diopside, potassic titanites (e.g. priderite), potassic zirconian silicates (e.g. wadeite), shcherbakovite, and armalcolite (cf. Contini et al. 1993). Lamproites characteristically do not contain plagioclase, nepheline, or melilite (Bergman 1987). Geochemically, lamproites have high K_2O/Al_2O_3 ratios (> 0.6; Foley et al. 1987), moderately high CaO (> 4 wt %) and very low Al_2O_3 contents (< 12 wt %; Mitchell and Bergman 1991). Most lamproites are peralkaline and have (Na+K)/Al ratios > 1 (Mitchell and Bergman 1991). Additionally, they are characterized by high concentrations of mantle-compatible elements (e.g. ~ 150 ppm V, ~ 400 ppm Cr, ~ 250 ppm Ni), high LILE (e.g. ~ 6000 ppm Ba, ~ 2000 ppm Sr), and high LREE (e.g. ~ 250 ppm La, ~ 400 ppm Ce; Mitchell and Bergman 1991).

In the previous literature, there has been some debate about whether to consider lamproites as a distinctive petrogenetic group, as proposed by Bergman (1987) and Mitchell and Bergman (1991), or to include them into the lamprophyre clan as suggested by Rock (1991). In hand specimen, lamproites appear very similar to rocks of the lamprophyre clan (e.g. the shoshonitic lamprophyre varieties kersantite and minette, and most alkaline lamprophyres; Müller et al. 1992a, 1993a) due to their porphyritic textures with abundant ferromagnesian phenocrysts such as phlogopite and lack of leucocratic phenocrysts. However, based on mineralogical and geochemical considerations, they are quite different. In contrast to lamprophyres, lamproites may contain alkali amphiboles such as riebeckite or richterite, but they lack plagioclase, a major component of many lamprophyres. Lamproites also have much lower SiO_2 (< 40 wt %) and Al_2O_3 (< 12 wt %) contents than shoshonitic or alkaline lamprophyres (commonly > 45 wt % and > 14 wt %, respectively; Rock 1991). Based on their exotic mineralogy, their distinct geochemistry, and their very rare occurrence in nature, lamproites are not further considered in this book.

2.5.3 Kamafugites

Kamafugites are mafic kalsilite-bearing lavas, and they represent the rarest examples of the magmatic rocks (Mitchell and Bergman 1991). Kamafugites may occur as dykes or lavas which are restricted to within-plate settings. Important type-localities are the igneous rocks (e.g. katungites, mafurites, ugandites) from the Toro Ankole region, Uganda (Holmes 1950; Barton 1979), and those from Cupaello and San Venanzo, Italy (Mittempergher 1965; Gallo et al. 1984). The term "kamafugite" (*ka*tungite-*ma*furite-*ug*andite) was introduced by Sahama (1974), and subsequently has been established in the modern literature (cf. Foley et al. 1987).

Mineralogically, kamafugites are characterized by the presence of olivine phenocrysts in a groundmass consisting of phlogopite, clinopyroxene, leucite, melilite, perovskite, and kalsilite, the latter reflecting their very low SiO_2 contents (Gallo et al. 1984; Foley et al. 1987). Kamafugites are petrographically distinguished from lamproites by the presence of kalsilite and melilite, and absence of sanidine (Mitchell and Bergman 1991). Apatite and perovskite normally represent only minor phases (Gallo et al. 1984). Kamafugites are geochemically distinct with extremely low SiO_2 (< 45 wt %), very low Al_2O_3 (< 12 wt %), low Na_2O (< 1.38 wt %), and very high CaO contents (> 8 wt %), as discussed by Gallo et al. (1984) and Foley et al. (1987). Their characteristically high concentrations of LREE (e.g. up to 470 ppm Ce) and high field-strength elements (HFSE) (e.g. up to 44 ppm Y, up to 680 ppm Zr; Gallo et al. 1984) are consistent with their restricted occurrence in within-plate settings.

2.5.4 Orogenic Ultrapotassic Rocks

The group of orogenic ultrapotassic rocks is equivalent (Foley et al. 1987) to the highly potassic igneous rocks from the Roman Magmatic Province, Italy (e.g. Holm et al. 1982; Rogers et al. 1985; and see Chap. 4).

Geochemically, orogenic ultrapotassic rocks are characterized by relatively low K_2O/Al_2O_3 ratios (< 0.5) when compared with the extreme K_2O-enrichments of lamproites and kamafugites (K_2O/Al_2O_3 > 0.6). They may occur either as dykes (e.g. Müller et al. 1993a) or as lavas (e.g. Cundari 1973). Orogenic ultrapotassic rocks from the Roman Province typically have high Al_2O_3 contents (> 12 wt %; Civetta et al. 1981; Holm et al. 1982; Rogers et al. 1985).

2.6 Group II Kimberlites

Kimberlites are rare, volatile-rich, ultrabasic, potassic igneous rocks occupying small vents, sills, and dykes (Dawson 1987). Kimberlites have been divided by Smith et al. (1985) into two distinct varieties termed Group I and Group II kimberlites. Group I is also known as mica-poor and Group II as mica-rich kimberlites (Dawson 1987).

Petrographically, Group I kimberlites are characterized by the presence of olivine, phlogopite, apatite, monticellite, calcite, serpentine, and minor magnesian ilmenite (Mitchell 1989). Groundmass spinels and perovskite are abundant (Mitchell 1989). Group II kimberlites, which have been only recognized in South Africa and Swaziland to date, are dominated by phlogopite, and minor diopside and apatite phenocrysts (Mitchell 1989). Their groundmass is mainly phlogopite, diopside, and leucite, whereas monticellite and magnesian ilmenite are absent (Skinner 1989). Group II kimberlites are rarely accompanied by other potassic intrusions such as lamprophyres (Dawson 1987). Both groups are geochemically different, with higher concentrations of P, Rb, Ba, and LREE, and lower concentrations of Ti and Nb in Group II kimberlites (Smith et al. 1985; Skinner 1989). Importantly, Group II kimberlites are highly potassic, with K_2O contents of about 3 wt % (cf. Mitchell 1989).

2.7   Potassic Igneous Rocks as Considered in this Book

Potassic igneous rocks, as considered in this book, are defined by molar K_2O/Na_2O ratios of about or slightly higher than unity (cf. Peccerillo 1992). While potassic igneous rocks from within-plate settings tend to occur as isolated geological bodies, those from subduction-related tectonic settings normally occur as the end-members of a continuous igneous-rock spectrum that might range from boninites and tholeiites to high-K calc-alkaline rocks and shoshonites during arc evolution (see Chap. 3). In these settings, the authors also consider rocks with molar K_2O/Na_2O ratios < 1, if the whole-rock compositions are K_2O > 1 wt % at about 50 wt % SiO_2 (e.g. the basalts from the Mariana Arc: see discussion in Chaps 3 and 4).

This book does not consider Group II kimberlites, due to their very distinctive geochemistry with low SiO_2 (about 36 wt %) and very high MgO contents (about 30 wt %), their exotic mineralogy, and their rare occurrence in nature (cf. Mitchell 1989). The book also excludes lamproites and kamafugites (Groups I and II of Foley et al. 1987), because of their limited occurrence in nature and their exotic mineralogy (e.g. rare minerals including richterite, melilite, perovskite, and priderite occur as well as leucite and/or kalsilite). They tend to be isolated and are normally not associated with other high-K rocks. Lamproites and kamafugites occur typically in mobile belts at craton margins (lamproites) or in rift valleys (kamafugites), but not in orogenic areas, and they are not associated with gold or base-metal mineralization. It is considered, therefore, that eliminating lamproites and kamafugites from the potassic igneous rock database SHOSH2 (database discussed in Chap. 3) not only has a sound basis, but also allows the discrimination to concentrate on finer chemical differences among the remaining *orogenic* potassic igneous rocks (including groups III and IV ultrapotassic rocks), which are of more interest here. This decision does leave some leucite-bearing rocks in SHOSH2, such as those of the Roman Province, Italy, and the leucitites of New South Wales, Australia. Lamproites and kamafugites were eliminated on a province-by-province basis, but the criteria are equivalent to such chemical screens as $CaO/Al_2O_3 < 1.3$, $CaO < (SiO_2 - 30)$, $CaO > (21 - SiO_2)$, and $CaO > (22 - 1.25 \cdot Al_2O_3)$, based on figures 1 and 3 of Foley et al. (1987).

2.8   Field Recognition of Potassic Igneous Rocks

There is no golden rule for the recognition of potassic igneous rocks in the field because the characteristics vary from more mafic to more felsic varieties, and from plutonic to volcanic settings.

Volcanic and hypabyssal high-K rocks are generally characterized by porphyritic textures with phenocrysts of clinopyroxene, amphibole, biotite, phlogopite, and apatite in a fine-grained groundmass which is dominated by plagioclase and orthoclase. Extrusive shoshonitic igneous rocks are commonly dominated by plagioclase and clinopyroxene phenocrysts which are accompanied by amphibole, mica, and apatite phenocrysts. However, the volatile-rich phenocrysts such as amphibole, biotite, phlogopite, and apatite are mainly developed in lamprophyres, which crystallize at shallow levels in the crust under low confining pressures of the overlying rocks.

Plutonic high-K rocks are generally characterized by equigranular textures comprising larger crystals of plagioclase, amphibole, biotite, and phlogopite in a medium-grained groundmass of orthoclase and plagioclase. Typical examples are the potassic plutons which host several major porphyry copper-molybdenum deposits in the USA. In mining company reports, they are commonly referred to as "monzonite-porphyries", or "diorites".

Potassic igneous rocks are commonly dark grey. However, they can be pink due to the presence of Fe in the orthoclase structure. In rare cases, mineralized high-K intrusions are overprinted by an intense hematite-sericite alteration, resulting in brick-red colours (e.g. Goonumbla and Cadia porphyry copper-gold systems, New South Wales, Australia).

3   Tectonic Settings of Potassic Igneous Rocks

3.1   Introduction

Modern potassic igneous rocks occur in a wide range of tectonic settings, from continental to oceanic and within-plate, some of which are not apparently associated with subduction (Joplin 1968; Morrison 1980; Müller et al. 1992b). It is therefore important, whether improving exploration models for ancient mineral deposits, or reconstructing ancient terranes, to be able to distinguish the tectonic settings in which ancient potassic igneous rocks were generated. The following chapter seeks to provide such a distinction.

3.2   Tectonic Settings of Potassic Igneous Rocks

Young (< 60 Ma) potassic igneous rocks have been recognized throughout the world in five principal tectonic settings (Fig. 3.1, Table 3.1), of which two are closely related (Müller et al. 1992b). A schematic overview of the different tectonic settings in which potassic igneous rocks occur is shown in Figure 3.2.

3.2.1   Continental Arc

Continental arc potassic igneous rocks are well represented in the Andean volcanic belt (e.g. Venturelli et al. 1978) and in the Aeolian Islands in the Mediterranean (Ellam et al. 1989). Such settings are associated with reorganization of plate boundaries due to oblique plate convergence, and are normally characterized by relatively flat subduction angles and broad Benioff Zones. The nature and distribution of magmatic activity in the overriding plate is a function of the convergence rate, the age of the subducted lithosphere, and the presence of features such as seamount chains or aseismic ridges (Wilson 1989).

Table 3.1.   Data sources for potassic igneous rocks in filtered database SHOSH2. 2/6 = 2 analyses were retained in filtered database SHOSH2 from 6 analyses in unfiltered database SHOSH1 and original reference. From Müller et al. (1992b).

1. Continental arcs	2. Postcollisional arcs	3a. Oceanic arcs (initial)
Aegean Islands, Greece [a]	*Alps (Eastern)*	*Mariana Islands* [a]
Pe-Piper (1980) 4/8	Deutsch (1984) 5/5	Bloomer et al. (1989) [b] 2/7
	Müller et al. (1992a) 8/11	Dixon and Batiza (1979) 6/10
Aeolian Islands, Italy		Garcia et al. (1979) 1/6
Ellam et al. (1989) 7/22	*Alps (Western)* [a]	Lin et al. (1989) [c] 2/7
Keller (1974) 3/9	Beccaluva et al. (1983) 6/6	Meijer and Reagan
	Dal Piaz et al. (1979) 2/13	(1981) 14/14
Andes, Chile	Venturelli et al. (1984) 5/7	Stern (1979) 2/3
Deruelle (1982) 1/5		Taylor et al. (1969) 1/2
Thorpe et al. (1976) 3/4	*Iran (Northeast)*	
	Spies et al. (1984) 4/6	
Andes, Peru [a]		
Kontak et al. (1986) 4/4	*Papua New Guinea*	
	BMR [d] (unpubl. data) 83/166	
North American Cordillera [a]	De Paolo and Johnson (1979) 1/8	
Costa Rica	Jakes and Smith (1970) 13/24	
Reagan and Gill (1989) 4/4	Jaques (1976) 3/17	
Mexico	McKenzie and Chappell	
Allan and Carmichael	(1972) 3/10	
(1984) 3/3	J.P. Richards (unpubl. data,	
I.S.E. Carmichael	1988) 16/30	
(pers. comm.) 2/2	Smith (1972) 2/29	
Luhr and Kyser (1989) 1/1	Sombroek (1985) 19/47	
Luhr et al. (1989) 7/8		
New Mexico, Rio Grande	*Roumania*	
Duncker et al. (1991) 2/8	Peccerillo and Taylor	
	(1976b) [c] 4/4	
Roman Province [a]		
Appleton (1972) 2/12		
Civetta et al. (1981) 6/6		
Cox et al. (1976) 2/4		
Cundari (1979) 6/14		
Cundari and Mattias (1974) 5/17		
Fornaseri et al. (1963) 7/37		
Ghiara and Lirer (1976) 2/6		
Holm et al. (1982) 2/6		
Poli et al. (1984) 4/4		
Rogers et al. (1985) 7/8		
Savelli (1967) 5/19		
Thompson (1977) 1/2		
van Bergen et al. (1983) 8/8		

[a] Additional references were incorporated in SHOSH1, but all analyses were filtered out for SHOSH2.
[b] Only major elements.
[c] Only trace elements.
[d] Bureau of Mineral Resources, Geology and Geophysics, Australia (now Australian Geological Survey Organisation), PETCHEM database.

3b. Oceanic arcs (late)	4. Within-plate settings

Fiji [a]
Gill and Whelan (1989) 1/9

Kuril Islands
Bailey et al. (1989) 3/4

Sunda Arc
Foden (1979) 12/121
J.D. Foden (unpubl. data)
  5/97
Foden and Varne (1980) 6/10
Hutchison and Jezek
  (1978) 6/19
Wheller (1986) 29/122
Whitford (1975) 12/160
Whitford and Jezek
  (1979) 2/8
Whitford et al. (1979) 2/13

Vanuatu[a]
Gorton (1977) 10/12

Borneo
Bergman et al. (1988) 6/11

Gough Island, Atlantic
Le Maitre (1962) 4/13
Le Roex (1985) 7/19
Weaver et al. (1987) 2/4

North American Cordillera [a]
Arizona
Nicholls (1969) 2/2
Roden (1981) 2/2
Roden and Smith (1979) 1/2
Rogers et al. (1982) 5/5
Colorado
Alibert et al. (1986) 1/1
Leat et al. (1988) 6/6
Thompson et al. (1984) 4/5
California, Sierra Nevada
Dodge and Moore (1981) 18/19
van Kooten (1980) 13/13
Wyoming
Barton and van Bergen
  (1981) [b] 1/5
Gest and McBirney (1979) 2/5
Nicholls and Carmichael
  (1969) 1/4

New South Wales, Australia
Cundari (1973) 32/37

Tristan da Cunha, Atlantic [a]
Weaver et al. (1987) 2/4

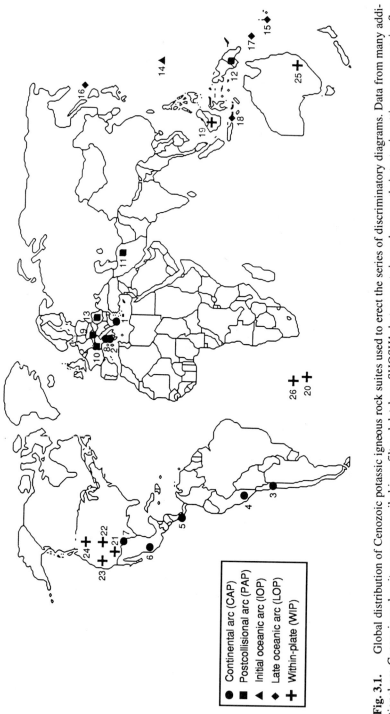

Fig. 3.1. Global distribution of Cenozoic potassic igneous rock suites used to erect the series of discriminatory diagrams. Data from many additional pre-Cenozoic rock suites were compiled in unfiltered database SHOSH1, but are not shown because their tectonic settings are uncertain. See Table 3.1 for data sources. 1 Aegean Islands, 2 Aeolian Islands, 3 Chile, 4 Peru, 5 Costa Rica, 6 Mexico, 7 New Mexico, 8 Roman Province, 9 Eastern Alps, 10 Western Alps, 11 Iran, 12 Papua New Guinea, 13 Roumania, 14 Mariana Islands, 15 Fiji, 16 Kuril Islands, 17 Vanuatu, 18 Sunda Arc, 19 Borneo, 20 Gough Island, 21 Arizona, 22 Colorado, 23 California, 24 Wyoming, 25 New South Wales, 26 Tristan da Cunha. From Müller et al. (1992b).

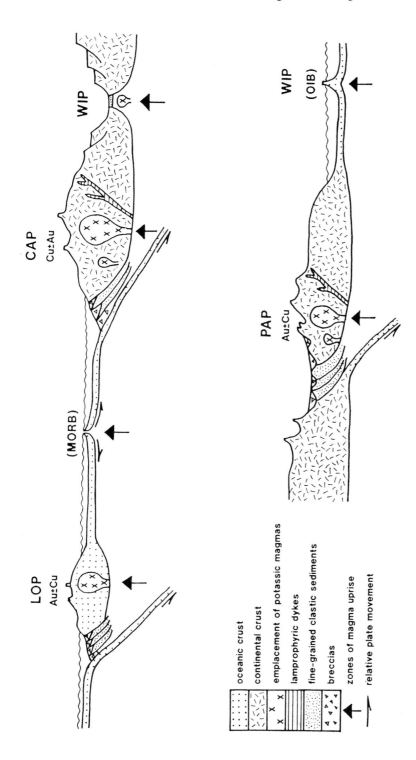

Fig. 3.2. Schematic overview of potassic igneous rocks from different tectonic settings. CAP = continental arc; PAP = postcollisional arc; LOP = late oceanic arc; WIP = within-plate setting; MORB = mid-ocean ridge basalt; OIB = oceanic island basalt. Modified after Mitchell and Garson (1981).

3.2.2 Postcollisional Arc

Postcollisional arc potassic igneous rocks are exemplified by the Eastern and Western Alps (e.g. Venturelli et al. 1984; Müller et al. 1992a), where the continental plates collided during the Eocene and subduction has long since ceased. This setting represents the most complex case of subduction-related magmatism, in which the suture zone forms an area of crustal thickening, characterized by complex magmatic activity and tectonic uplift (Wilson 1989). After collision, potassic igneous rocks may be emplaced as dykes, commonly followed by alkaline volcanism where extensional tectonic regimes develop as a consequence of uplift (Wilson 1989; Müller et al. 1992a).

3.2.3 Oceanic (Island) Arc

Oceanic (island) arc potassic igneous rocks are generated at the site of subduction of one oceanic lithospheric plate beneath another. Oceanic arc settings normally show steep subduction angles and, compared to continental arc settings, relatively short distances between volcanic arc and subduction trench where projected to surface. High-K rocks from this setting can be subdivided into two types: *initial* and *late* oceanic arc potassic igneous rocks.

Initial Oceanic Arc Potassic Igneous Rocks. Initial oceanic settings are exemplified by unusual potassic igneous rocks (including shoshonites) from the northern Mariana Arc (Stern et al. 1988; Bloomer et al. 1989; Lin et al. 1989). Whereas the initial and fore-arc melts in most island arcs have boninitic or low-K tholeiitic affinities, and potassic igneous rocks occur only in the mature, back-arc stages of arc evolution, the Mariana potassic igneous rocks occur along the magmatic front, and may represent the reconstruction of the arc following an episode of back-arc rifting (Stern et al. 1988).

Late Oceanic Arc Potassic Igneous Rocks. Late oceanic arc settings are well represented in Fiji (Gill 1970) and the New Hebrides (Gorton 1977; Marcelot et al. 1983) in the western Pacific. Here, potassic igneous rocks form the *youngest* volcanic products; they were erupted after lower K-tholeiitic and/or calc-alkaline rocks and farthest from the trench, in the classic volcanic sequence referred to above.

3.2.4 Within-Plate

Within-plate potassic igneous rocks are not related to any form of subduction. They are particularly well represented in the North American Cordillera (Fig. 3.1, Table 3.1). They may be associated with hot-spot activity or with extensional (particularly rift) tectonics (e.g. in the western branch of the East African Rift), and the magmas from which they crystallize are commonly generated at greater depths than the other four categories.

3.2.5 Problems with Tectonic Classification

In areas of plate tectonic complexity, there may be ambiguity about the classification even of young potassic igneous rocks into one of these five settings. In the western Sunda Arc of Indonesia (Curray et al. 1977), for example, not only does the crustal seismic-velocity structure change from continental to oceanic moving east from Sumatra into Java, but the subduction angle and traceable depth of subduction also change from oblique, shallow, and < 200 km in Sumatra to orthogonal, steep, and > 600 km in Java. Reflecting these changes, potassic igneous rocks occur on the back-arc side of Java and islands further east (Whitford et al. 1979), and are hence attributed to a late oceanic arc setting. In Sumatra, however, they are relatively early, occur on the fore-arc side (Rock et al. 1982), and are better attributed to an initial oceanic arc setting.

Other difficulties arise in continental-scale igneous provinces such as the Cenozoic of the North American Cordillera. Although potassic igneous rocks in young volcanic suites along the western seaboard (e.g. Crater Lake, Oregon; Bacon 1990) are unequivocally continental arc, and potassic igneous rocks from states well inland (e.g. North Dakota; Kirchner 1979) are within-plate by definition, many potassic igneous rocks from intervening states (e.g. Colorado, Wyoming; Gest and McBirney 1979; Leat et al. 1988) could be of either affinity. Following most previous authors, within-plate affinities are assumed where there is a clear association with rifting (e.g. Sierra Nevada lavas, California; van Kooten 1980), or where the depth of magma generation is too great for subduction affinities (e.g. Navajo Province, New Mexico; Rock 1991).

A further set of ambiguities arises from differing opinions in the literature. For example, Cundari (1979) uniquely assumes a within-plate setting for the Sabatini lavas of Italy, whereas all other authors concerned with Roman Province potassic igneous rocks assume a continental arc setting (e.g. Civetta et al. 1981).

3.3   History of Discrimination of Tectonic Setting by Geochemical Means

First put forward to distinguish the tectonic settings of ancient basalts (e.g. Pearce and Cann 1973), geochemical discrimination diagrams were used widely in the late 1970s and early 1980s. Equivalent diagrams were introduced to classify granites (sensu lato) tectonically (e.g. Pitcher 1983; Pearce et al. 1984), and the method was also extended to the petrological classification of altered and/or metamorphosed ig-

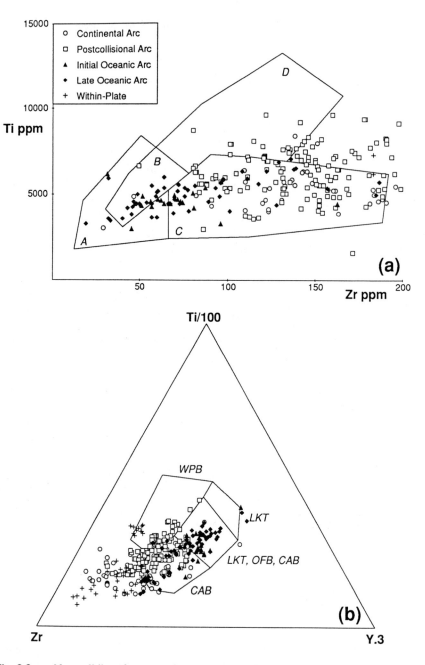

Fig. 3.3.   Non-validity of two popular geochemical discrimination diagrams (Pearce and Cann 1973) when applied to potassic igneous rocks. Data from SHOSH2. In (a), fields A + B = low-K tholeiites (LKT), B + C = calc-alkaline basalts, B + D = ocean-floor basalt. In (b), WPB = within-plate basalts, LKT = low-K tholeiites, OFB = ocean-floor basalts, CAB = calc-alkaline basalts. From Müller et al. (1992b).

neous rocks (e.g. Floyd and Winchester 1975). As with many other geological methods, a period of criticism and reappraisal followed (e.g. Smith and Smith 1976; Morrison 1978), but these drawbacks have proved insufficient to limit the use of the method. Geochemical discrimination diagrams, although initially empirical, subsequently received a formal basis from statistical analysis and theoretical arguments (e.g. Pearce 1976).

The principle behind the successful use of these diagrams is the delineation of trace-element differences between modern rocks in different known settings, based on a comprehensive database; these differences are then depicted in diagrams which can be used to assign older samples from equivocal tectonic settings. The assumed relatively immobile HFSE (namely Ti, P, Y, Zr, Nb, Hf, Ta, Th, REE), are generally considered to be most suitable for use in these diagrams, although some studies have suggested that Th (Wood et al. 1979; Villemant et al. 1993) and REE (Hellman et al. 1979) may be mobile under certain conditions, and most of these elements may be somewhat mobile in highly altered, mineralized wallrocks.

Previous discrimination diagrams, which have been developed for basalts and granitic rocks, are not suitable for discriminating the tectonic setting of potassic igneous rocks. For example:

− Ti-Zr and Ti-Zr-Y diagrams of Pearce and Cann (1973): Potassic igneous rocks extend to compositions well outside the defined fields on these diagrams and, in particular, within-plate potassic igneous rocks normally show much higher Zr concentrations than indicated by the defined within-plate field for other igneous rocks (Fig. 3.3).
− Ti-Zr-Sr diagram of Pearce and Cann (1973): This diagram cannot separate within-plate from subduction-related potassic igneous rocks. Most potassic igneous rocks plot misleadingly in the calc-alkaline field, and those from postcollisional settings plot erroneously into the ocean-floor basalt field. The discrimination is also subject to the severe limitation of Sr mobility for altered rocks.
− Zr/Y versus Zr diagram of Pearce and Norry (1979): Subduction-related potassic igneous rocks from continental arcs and from intra-oceanic tectonic settings plot erroneously within the mid-ocean ridge basalt (MORB) and within-plate basalt fields on this diagram.
− Hf/3-Th-Ta diagram of Wood et al. (1979): Nearly all potassic igneous rocks, even those from known within-plate tectonic settings, plot misleadingly into the subduction-related field on this diagram.

The only previously developed diagrams that actually accommodate potassic igneous rocks (Pearce 1982), only allow them to be identified petrologically; the diagrams do not discriminate the tectonic settings of the rocks (see Fig. 3.4).

Another common plot for comparing geochemical patterns is the spidergram (Thompson 1982), but spidergrams do not effectively separate potassic igneous rocks from different tectonic settings (Fig. 3.5). For example, although arc-related potassic igneous rocks (Fig. 3.5a–d) show relatively high values of K, Rb, Cs, Ba, and

Pb (Sun and McDonough 1989) and the supposedly diagnostic (negative) Ti-Nb-Ta (TNT) anomalies (Saunders et al. 1980; Briqueu et al. 1984; Foley and Wheller 1990), these features are also shown by some within-plate potassic igneous rocks (e.g. the potassic lamprophyres from Borneo; Bergman et al. 1988; see Fig. 3.5e). There is no simple relationship between TNT anomalies in spidergram patterns and subduction-related processes of magma generation, because potassic igneous rocks without those anomalies can also occur in subduction settings (Rock 1991).

Many previous workers have tried to explain the negative anomalies of the HFSE Ti, Nb, and Ta, in terms of their retention in the subducted oceanic slab during the dehydration process. This is because these elements are thought to have low solubilities in the metasomatic fluids that transport mantle-incompatible elements from the subducted slab into the overlying mantle wedge (e.g. Saunders et al. 1991). It has been suggested also that these elements were retained in the mantle wedge in minerals such as rutile, ilmenite, and titanite during partial melting (e.g. Foley and Wheller 1991). However, more recent studies attempt to explain the Ti depletion of arc magmas as a result of the higher fO_2 in subduction zones. Higher temperatures are required to melt Ti-bearing phases when fO_2 is higher, thus producing Ti depletion in the potassic melts derived from subduction zones (Edwards et al. 1994). Other studies suggest that the HFSE depletions in arc magmas are caused by the precipitation of HFSE-bearing phases as the melt migrates upwards through the mantle wedge, while simultaneously dissolving phases with relatively low HFSE abundances (Kelemen et al. 1990; Woodhead et al. 1993).

More recently, the results of Ionov and Hofmann (1995) suggest that amphibole is an important host mineral for Nb and Ta in the upper mantle, and may control the development of negative Nb-Ta anomalies in arc magmas. Ionov and Hofmann (1995)

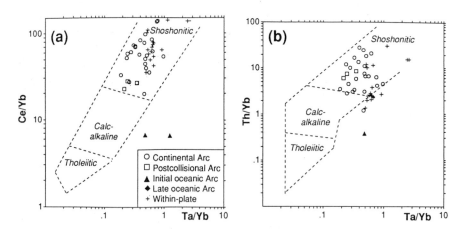

Fig. 3.4. Confirmation of shoshonitic affinities of rocks used in this study. Diagram from Pearce (1982), data from SHOSH2. Note that only a minority of samples in SHOSH2 have determinations for Ta, Yb, Ce and/or Th and can thus be plotted on this diagram. From Müller et al. (1992b).

postulated a model in which fluids, generated by dehydration of the subducted slab, ascend through the mantle wedge and precipitate amphiboles. Niobium and Ta are transferred with these fluids into the mantle wedge, where these elements are partitioned into crystallizing amphibole, thus inducing low Nb and Ta concentrations in the residual fluid. As the residual fluids migrate further, they may induce partial melting in high-temperature regions of the mantle wedge, thus producing melts with negative Nb and Ta anomalies (Ionov and Hofmann 1995).

In view of the discussions above, the use of geochemical discrimination diagrams in isolation from other lines of evidence for tectonic settings is never recommended. Wherever possible, geochemistry should be combined with other geoscientific information, and attempts to combine all these lines of evidence into an expert-system should be made (Pearce 1987). However, understanding of potassic igneous rocks is at present insufficiently advanced for such a sophisticated approach. The pilot study presented here, therefore, attempts to show that potassic igneous rocks from different tectonic settings are geochemically distinct, and then uses this to erect discrimination diagrams for the tectonic settings of older potassic igneous rocks.

3.4   Erection of Databases SHOSH1 and SHOSH2

There are many possible options to balance the breadth of the database used against the precision of discrimination actually achieved. For example, at one extreme, it might be possible to discriminate the tectonic setting of *all* igneous rocks irrespective of their compositions; this would be comprehensive in its scope, but limited in its discriminatory power. At the other extreme, it might be possible to restrict attention to a very narrow range of compositions (e.g. rocks within the shoshonite field on Fig. 3.6b); this would no doubt achieve much better discrimination, but would be very limited in its scope. In general, the more diffuse (less internally coherent) the database, the lower its discriminatory power.

There is also a multitude of options for screening an initial database. Many authors have used chemical screens: for example, Pearce and Cann (1973) restrict attention to analyses with total (MgO + CaO) contents between 12 and 20 wt %, and Pearce and Norry (1979) use those analyses with total alkalis below 20 wt %. However, such screens are always arbitrary and artificially imposed. It is considered preferable here to embrace the entire natural compositional range of an igneous suite as far as possible. The major exception, where the argument for screening is irrefutable, is in isolating chemical differences that are due to tectonic setting (that is, reflect the nature of the source region and magma generation processes within it), from those that reflect the subsequent history of the magma (magmatic differentiation or accumulation, secondary alteration or weathering, etc.) This usually requires restricting the rocks within a database to fresh and primitive samples.

This book attempts a compromise in both the breadth of the database used and in its screening so that it is sufficiently broad in scope but still provides critical dis-

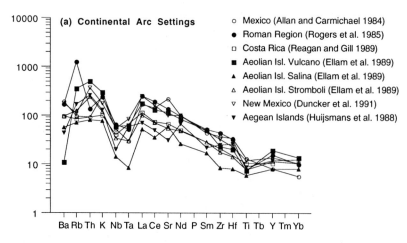

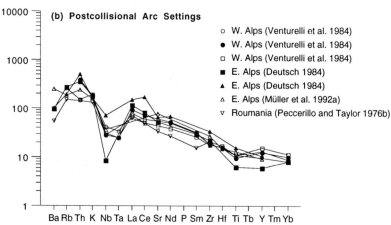

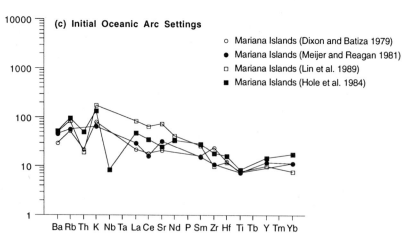

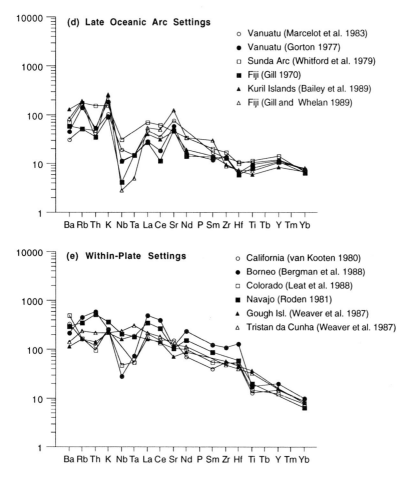

Fig. 3.5.   Representative chondrite-normalized spidergram patterns for potassic igneous rocks from the five tectonic settings recognized in this study. Element order and normalizing factors after Thompson (1982). Sources are listed in references. From Müller et al. (1992b).

criminatory power. Over 100 published references containing relevant data were first identified by combining traditional (manual) and computerized literature search, using keywords such as "shoshonitic", "K-rich", "high-K", and "potassic". Bibliographic indexes searched include *Mineralogical Abstracts*, *Geological Abstracts* and the CD-ROM version of *GeoRef*. Some large existing source databases such as IGBA, LAMPDA (Rock 1991), the ultrapotassic rocks database of Foley et al. (1987), and PETCHEM (Australian Bureau of Mineral Resources) were also searched. Data from all these sources were supplemented by 50 new high-precision analyses of potassic igneous rocks from Australia and Papua New Guinea for a comprehensive suite of up to 35 major and trace elements. Methods outlined by Rock (1988, 1991) for the

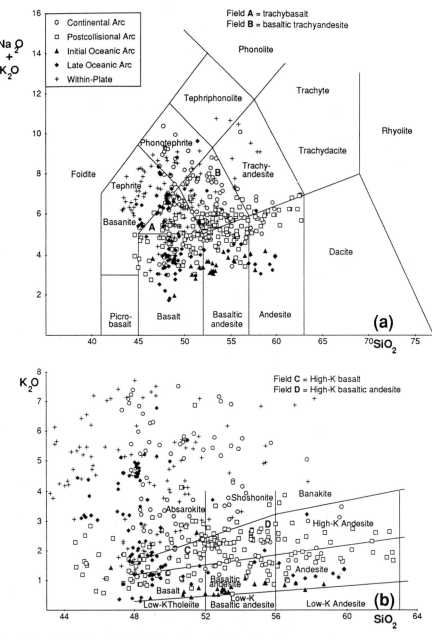

Fig. 3.6. Established classification diagrams illustrating the range of compositions in the filtered database SHOSH2. (a) TAS diagram recommended by the IUGS Subcommission on Igneous Rock Systematics (Le Maitre, 1989). (b) K_2O-SiO_2 diagram (Peccerillo and Taylor 1976a) now widely adopted in the literature; this differs in essence from the equivalent IUGS diagram only in the absence of the top three fields. Analyses have been recalculated to 100 % free of volatiles before plotting as wt % in both diagrams. From Müller et al. (1992b).

compilation of petrological databases were used to validate and check the quality of the data, to classify them consistently, and to eliminate duplicates. Altogether, this yielded an initial global database (SHOSH1) comprising 2,222 analyses of potassic igneous rocks for up to 11 major and 24 trace elements, all classified according to igneous province, occurrence (i.e. suite or volcano), age, and tectonic setting. Authors' published descriptions and classifications of these features were adhered to, except in cases of significant inconsistency. SHOSH1 includes many relatively K-poor (calc-alkaline) compositions coeval with potassic igneous rocks, but is not intended to cover the whole spectrum of orogenic volcanic rocks from low-K to high-K.

SHOSH1 was then carefully filtered and checked as follows to generate a final working database (SHOSH2) of 497 analyses (for the same range of 35 elements):

1. Age: All pre-Cenozoic analyses were eliminated.
2. Alteration: All analyses with > 5 wt % loss-on-ignition (LOI) were eliminated. This limit was not entirely arbitrary, but marked a natural break in the rocks in SHOSH1 between apparently fresh and more weathered or altered samples.
3. Primitive Chemistry: To eliminate evolved and cumulate samples, as mentioned above, all analyses with mg# outside the range 0.5–0.8 were filtered out.
4. Potassic versus Ultrapotassic: A more complex and subjective decision involved whether to retain ultrapotassic as well as potassic rocks. Ultrapotassic rocks commonly contain leucite, whereas potassic rocks, such as shoshonites, normally do not. Fortunately, an exhaustive global survey of these already exists (see Foley et al. 1987), and their definitions are outlined in Chapter 2. Although orogenic ultrapotassic rocks are considered in this study, the ultrapotassic lamproites and kamafugites have been excluded, as discussed in Chapter 2.
5. Outliers: Outlying compositions not only reduce the internal coherence of a database (and hence reduce the potential efficiency of discrimination), but are quite likely to be samples which are altered, weathered, or otherwise unrepresentative. The analyses which remained after stages 1–4 were therefore plotted on various standard classification diagrams (e.g. Figs 3.4, 3.6), in order to eliminate gross outliers. For example, two analyses lying in the trachyte and dacite fields on Figure 3.6a and b were eliminated, not only on grounds of internal self-consistency, but also because Hf and Zr, which were expected to be useful discriminants in this study, may be lost via zircon fractionation from rocks with > 68 wt % SiO_2 (Pearce et al. 1984). Three analyses in the foidite field were also eliminated.
6. Classification: To ensure logical coherence, it was checked that the final contents of SHOSH2 were substantially potassic (Table 3.2), and that the minority of samples analyzed for the appropriate elements plotted predominantly in the shoshonitic fields on Figure 3.4.

The overall major-element spectrum of the analyses in SHOSH2, and the range of compositions to which the diagrams developed in the following sections therefore apply, are given at the top of Figure 3.7. The global distribution and numbers of

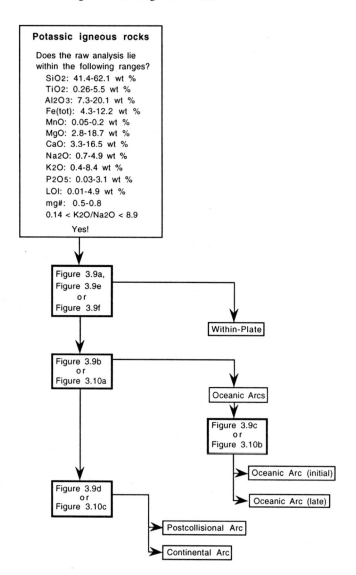

Fig. 3.7.   Suggested flow-chart for separating unknown samples of potassic igneous rocks into the five investigated tectonic settings. In order to achieve a maximum discrimination effect, the following scheme is recommended:

1. Plot unknown samples on Figures 3.9a, e and/or f to separate within-plate potassic igneous rocks.
2. Plot non-within-plate samples on Figures 3.9b or 3.10a to separate oceanic arc rocks.
3. Plot oceanic arc potassic igneous rocks on Figures 3.9c and/or 3.10b to separate those from initial and late settings.
4. Plot remaining samples on Figures 3.9d and/or 3.10c to discriminate between continental and postcollisional arc potassic igneous rocks.

From Müller et al. (1992b).

analyses from various suites are quoted in Table 3.1.

Although the above compilation and screening procedures are claimed to have been as careful and as scientifically thorough as the current literature permits, the resultant databases SHOSH1 and SHOSH2 are still recognized as suffering from a number of drawbacks, including the following:

– Homogeneity: Different published papers quote data for completely different sets of trace elements, and some include only major or only trace elements, hence both SHOSH1 and SHOSH2 are unavoidably replete with missing data for the full set of 35 elements compiled. Some elements such as Mo, Sb, Sn, and Cs were not compiled at all, because data were so few. Some published papers also fail to distinguish clearly between data which are "not analyzed" (missing) or "not detected", both being variously indicated by "nd" or "zero" in tables. It has been consistently assumed that these indicate the absence of data.

– Analytical Differences: It is impossible to make any allowance in a compilation such as this for the differing precision and accuracy of the wide range of analytical methods used in the literature (XRF, ICPMS, INAA, AAS, etc.), and particularly the lower precision of earlier data produced before the advent of the flame photometer. Although this is unlikely to be a problem for major elements, where at least the internal check of 100 wt % analytical totals is available, it is more of a problem for trace elements, on which tectonic setting discrimination for ancient potassic igneous rocks must substantially depend.

– Potassic Alteration: Potassic alteration is a widely recognized phenomenon in mineralized igneous systems, including porphyry copper and mesothermal gold deposits. For certain gold-associated potassic igneous rocks compiled in SHOSH1, it can be shown that some high K_2O values are not primary but alteration-induced (e.g. Porgera; Rock and Finlayson 1990). Although screening steps 2, 3, and 5 above are likely to have minimized this problem, there are simply insufficient descriptive (e.g. petrographic) data in most source references to determine the extent of alteration (and weathering) in individual samples incorporated into SHOSH1. Consequently, SHOSH2 may contain some analyses whose high K contents are at least partly due to secondary processes rather than primary enrichment.

– Initial Oceanic Arc Setting: Further difficulties are presented by the initial oceanic arc setting and whether it should be included at all. There is only one example of this suite (the Mariana Arc), which raises questions about its representativeness. Moreover, most available analyses from this one suite were eliminated by the above screening procedures, whereupon remaining data in SHOSH2 then included only one potassic rock (Table 3.2) and plotted outside the shoshonite field on Figure 3.4. It was eventually decided to retain the setting on the grounds that the unscreened data-set in SHOSH1 shows clear potassic affinities — one of 51 analyses lies in the absarokite and three in the TAS high-K basaltic andesite field on Figure 3.6b, and all five analyses in the trachyandesite field on Figure 3.6a are potassic. Nevertheless, the efficiency of discrimination for this particular setting is recognized as being limited.

Table 3.2.   Numbers of sodic, potassic, and ultrapotassic rock analyses included in final filtered database SHOSH2. From Müller et al. (1992b).

Tectonic setting	Sodic [a]	Potassic [a]	Unassigned [a]	Ultrapotassic [b]	Total [c]
Continental arc	0	60	36	42	96
Postcollisional arc	0	17	153	2	170
Initial oceanic arc	0	1	25	0	26
Late oceanic arc	0	47	42	20	89
Within plate	0	85	26	54	111
Totals	0	209	285	118	492

[a] General definition of IUGS (Le Maitre 1989): "sodic" means (Na_2O - 4 wt %) > K_2O, "potassic" means $Na_2O < K_2O$; otherwise unassigned.
[b] Definition of Foley et al. (1987): K_2O > 3 wt %, MgO > 3 wt %, and K_2O/Na_2O > 2.
[c] Sum of sodic, potassic and unassigned analyses; five analyses of the 497 in SOSH2 do not have values for Na_2O.

The ideal approach to the present problem would be to generate a more self-consistent database. This would comprise *new* data for fresh samples chosen carefully and at first hand, from intimate geological and petrographical knowledge of all the studied suites of rocks, and analyzed for exactly the same set of critical elements by the same laboratory and analytical techniques. Unfortunately, to generate such a database of the size of SHOSH1 (2222 analyses) is not a short-term option. The present work can therefore only be regarded as a pilot study, pending the availability of a database which approaches this ideal.

3.5   Discrimination of Tectonic Setting by Multivariate Statistical Methods

Multigroup linear discriminant analysis is the statistical method that optimizes separation between several groups of multivariate data. Le Maitre (1982) and Rock (1988) provide details of the method in a geological and specifically petrological context, and Pearce (1976) gives a particularly relevant application in separating basalts from different tectonic settings. Multigroup linear discriminant analysis does so by maximising the ratio of between-groups to within-groups variances. The raw data are first recast into a set of *discriminant functions*, namely linear weighted combinations of the measured major and/or trace element variables. There are N-1 discriminant functions for N groups of data, and for viability there should be at least NxM rock analyses (samples) in the database, where M is the number of variables (major and trace elements) to be used in the discriminant functions. Attempting to discriminate five tectonic settings will, therefore, lead to four discriminant functions, and the total of 497 potassic igneous rock samples in SHOSH2 is sufficient to allow any combination of the up to 35 measured major and trace elements to be used as discriminating variables. Figure 3.6 shows two established classification diagrams illustrating the range of compositions in the filtered database SHOSH2.

The main problem in applying multigroup linear discriminant analysis to SHOSH2 arises from the missing data referred to in Section 3.4. Multigroup linear discriminant analysis requires a complete and consistent data-matrix: i.e. with a measured value in every sample for every element to be incorporated in the discriminant functions. To generate such a data-matrix from SHOSH2, samples or elements with missing values must be eliminated, and obviously there is no unique solution to this process. Table 3.3 therefore presents the results of two end-member runs of multigroup linear discriminant analysis, one using major elements, for which 486 of the 497 incorporated potassic volcanic rocks have been analyzed (Table 3.3a); and one based on the optimum (i.e. largest possible) combination of samples and immobile elements (Table 3.3b).

Initial results showed that multigroup linear discriminant analysis can distinguish potassic igneous rocks from different tectonic settings, but with varying degrees of efficiency. Careful examination then identified several complete sets of analyses which were consistently misclassified. One of these was Cundari's (1979) analyses for the Sabatini lavas, assigned by him as within-plate; these were all classified by the multigroup linear discriminant analysis as continental arc, the setting assumed by all other authors (e.g. Civetta et al. 1981) concerned with the Roman potassic igneous rocks, and their assignment was therefore adjusted in SHOSH2. Other datasets which split between two or more tectonic settings were predictably those for which ambiguities were already known to apply, for example:

– Potassic volcanic rocks of Sierra Nevada (California) and parts of Wyoming were split between the within-plate and continental arc settings, either of which are perfectly admissible interpretations of their tectonic setting.
– Potassic igneous rocks from Chile and Peru were misclassified as postcollisional, possibly because they are sited away from the trench and may have ascended through thick continental crust (Thorpe et al. 1976; Dostal et al. 1977a, 1977b; Kontak et al. 1986).
– Potassic igneous rocks from Costa Rica were also misclassified as postcollisional, but were already known to be geochemically anomalous (Reagan and Gill 1989).

All these observations, far from undermining the basis of the treatment, are therefore considered to confirm very clearly that the multigroup linear discriminant analysis defines real compositional differences.

The final results of the multigroup linear discriminant analysis, after reassignment of these initially misclassified analyses, are summarized in Table 3.3a and Figure 3.8. In summary, 54% of continental arc, 100% of initial oceanic arc, 67% of late oceanic arc, 91% of postcollisional arc, and 79% of within-plate potassic igneous

Table 3.3.   Geochemical differences between potassic igneous rocks in five tectonic settings, revealed by multigroup discriminant analysis. CAP = continental arcs, PAP = postcollisional arcs, IOP = initial oceanic arcs, LOP = late oceanic arcs, WIP = within-plate settings.

(a) Based on major elements (486 samples divided among five tectonic settings)[a]

Dependent variable canonical coefficients (standardized by within-group pooled standard deviations)

Discriminant function	1	2	3	4
TiO_2	0.772	0.061	0.237	0.788
Al_2O_3	0.092	0.050	0.804	0.352
Fe_2O_3	-0.407	-0.211	-0.943	-0.394
MgO	0.511	0.136	0.904	-0.756
CaO	-0.270	-0.558	0.248	0.084
Na_2O	0.061	0.293	0.230	-0.916
K_2O	0.298	-0.654	0.690	-0.685
P_2O_5	0.270	0.235	-0.373	0.209

Prediction success: observed settings (rows) by predicted settings (columns) [b]

				Predicted			Observed
		CAP	PAP	IOP	LOP	WIP	totals
	CAP	**52**	19	6	8	11	96
	PAP	1	**154**	6	7	1	169
Observed	IOP	0	0	**21**	0	0	21
	LOP	11	2	15	**61**	0	89
	WIP	8	13	2	0	**88**	111
Predicted totals		72	188	50	76	100	486

[a] Both SiO_2 and LOI are omitted from this table to minimize the closure problem (Le Maitre 1982; Rock 1988). A check was also made to ensure that inclusion of SiO_2 did not significantly affect the classification efficiency.
[b] In prediction success tables, bold figures indicate analyses that are correctly assigned to their tectonic setting by the discriminant functions; e.g. in Table (b), 18 of 39 CAP analyses are correctly assigned by their chemistry whereas 10 are misassigned to the LOP setting.
[c] This table is a compromise between excessively limiting the number of samples and elements incorporated into the multigroup discriminant analysis; other important immobile elements (Hf, Ta, Th) were excluded outright because data were insufficient. Only four tectonic settings can be distinguished because no potassic igneous rocks from IOP settings (i.e. Mariana Islands) have been analyzed for the required element combination.

rocks can be correctly attributed to their tectonic setting from their major-element chemistry alone. In more detail, the minimal misclassification of initial oceanic arc samples indicates that these are very distinctive compositions, as evident from Figure 3.6b. Conversely, the relatively large misclassification of continental arc with respect to postcollisional arc and within-plate samples indicates that the three groups are similar. In Table 3.3b, the percentages of correctly classified analyses are predictably lower, because fewer variables have been employed on an unavoidably smaller data-set, but the pattern is similar. Table 3.3a is of interest in pinpointing geochemical differences attributable to a tectonic setting in young potassic igneous

(b) Based on maximum samples for immobile elements (150 samples divided among four tectonic settings)[c]

Dependent variable canonical coefficients (standardized by within-group pooled standard deviations)

Discriminant function	1	2	3
TiO_2	-0.670	0.598	-0.656
P_2O_5	0.409	0.406	0.057
Y	0.074	0.528	0.677
Zr	-0.779	-0.755	0.201
Nb	-0.181	-0.093	0.682
Ce	0.125	-0.543	-0.209

Prediction success: observed settings (rows) by predicted settings (columns) [b]

		Predicted				Observed totals
		CAP	PAP	LOP	WIP	
	CAP	**18**	7	10	4	39
Observed	PAP	6	**31**	4	0	41
	LOP	0	23	**26**	0	49
	WIP	7	1	0	**13**	21
Predicted totals		31	62	40	17	150

rocks not affected by potassic alteration, and hence when seeking genetic explanations. Table 3.3b is likely to be applicable to ancient potassic igneous rocks, where alteration and metamorphism effects are more likely, because it relies on immobile elements alone.

3.6 Discrimination via Simple Geochemical Diagrams

Since Figure 3.8 and Table 3.3 represent the best possible separation between the five groups that can be achieved on a two-dimensional diagram or by multivariate calculation, it is clear that no *single* graphical or mathematical treatment is likely to be adequate to assign an individual potassic igneous rock analysis to its tectonic setting. Any multigroup linear discriminant analysis treatment which includes all five settings will inevitably dissipate much of its power separating settings which are most distinctive, in turn leaving relatively little discriminatory power to separate the remainder. A *hierarchical* discrimination scheme is therefore appropriate, in which the most distinctive settings are successively stripped out, allowing each new discriminatory criterion in turn to target those subtle distinctions to which it is most suited. Since multigroup linear discriminant analysis is a relatively complex mathematical technique, and because its requirement of a full data-matrix is commonly impossible to meet (wherever literature data are involved), a hierarchical set of diagrams based on more conventional bivariate and triangular plots, rather than multivariate plots, was developed (Müller et al. 1992b).

The scheme is based as much as possible on immobile element *ratios,* including triangular plots, rather than absolute values, since such ratios are not only less affected by inter-laboratory variations, and easier to measure accurately, but are also theoretically unaffected by simple dilution or concentration affects such as the addition or removal of $CO_2 \pm H_2O$ during weathering, metamorphism and/or hydrothermal alteration. Most of the elements used in the developed diagrams are incompatible as well as immobile, so that their ratios are also little affected by fractionation or accumulation of major rock-forming minerals, and hence reflect primary source differences, such as those due to tectonic setting.

A first set of diagrams (Fig. 3.9) was based on the multigroup linear discriminant analysis itself, by using ratios of elements that have highest absolute but opposite canonical coefficients in Table 3.3. Adequate separation could not be achieved in some cases using simple ratios of immobile elements alone, but improved markedly when Al_2O_3 was used as a normalizing factor; since Al_2O_3 is the least mobile of the major elements, this result is considered to be reasonably satisfactory. Simple ratio

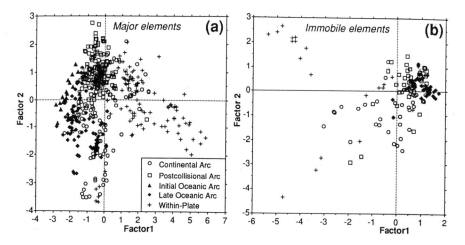

Fig. 3.8.   Discrimination diagrams for potassic igneous rocks from different tectonic settings, based on multigroup linear discriminant analysis using: (a) major-elements, (b) immobile elements only. The X and Y axes respectively plot the first two of the discriminant functions (canonical vectors or factors) which contribute most to the multidimensional group separation. Standardized versions of the weights (canonical coefficients) used to calculate these discriminant functions are given in the first two columns of Table 3.3, respectively. From Müller et al. (1992b).

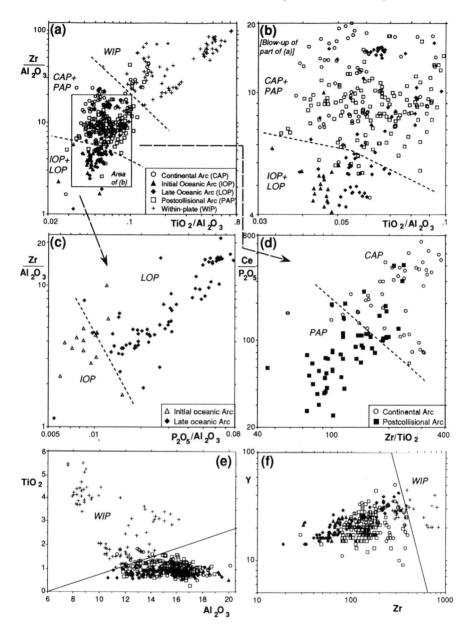

Fig. 3.9. Hierarchical set of discrimination diagrams for potassic igneous rocks from different tectonic settings, based on simple ratios of "immobile" elements revealed by discriminant analysis (Table 3.3) as contributing most effectively to group separation. As illustrated further in Figure 3.7, Figure 3.9a should be used first to extract within-plate potassic igneous rocks (Figs 3.9e and/or 3.9f are alternatives); Figure 3.9b should then be used to separate oceanic arc from continental and postcollisional arc settings, with the former distinguished as initial or late using Figure 3.9c, and the latter using Figure 3.9d. From Müller et al. (1992b).

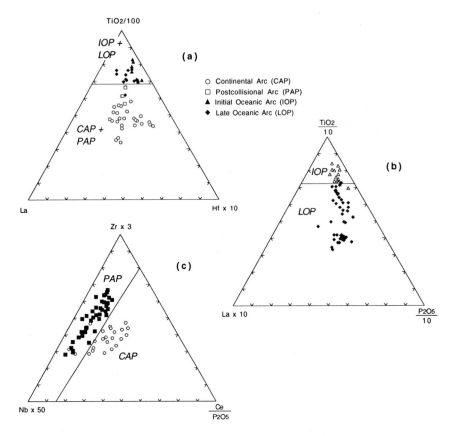

Fig. 3.10.   Discrimination diagrams for potassic igneous rocks from different tectonic settings, partly based on the more exotic trace elements. Numbers of points are much less than on Figure 3.9 owing to missing data. Figure 3.10a complements Figure 3.9b in separating continental and postcollisional arcs from oceanic arcs; Figure 3.10b complements Figure 3.9c in separating initial and late oceanic arcs; Figure 3.10c complements Figure 3.9d in separating continental and postcollisional arcs. The ratio TiO₂/100 in Figure 3.10a and the ratios TiO₂/10 and P₂O₅/10 in Figure 3.10b are calculated in ppm. From Müller et al. (1992b).

diagrams (e.g. Fig. 3.9) can never achieve the same separation efficiency as Figure 3.8 (or Table 3.3), because they necessarily use less of the total multivariate information in the data, yet they are more accessible for routine use. A second set of diagrams (Fig. 3.10) was devised more-or-less empirically, to take account of the more exotic trace elements, notably La, Ce, and Hf, which are only available for a minority of analyses in SHOSH2.

Together, these analyses lead to the hierarchical scheme outlined in Figure 3.7. In the first step, within-plate potassic igneous rocks can be separated from the four arc-related settings, by plotting data on Figures 3.9a, 3.9e, or 3.9f; samples with TiO₂ contents above 1.5 wt %, Zr above 350 ppm, or Hf above 10 ppm can be considered with particular confidence as within-plate types. In the second step, remaining samples should be plotted on Figure 3.10a, which discriminates oceanic arc from continental and postcollisional arc settings with almost 100% efficiency, based on the lower La and Hf contents of the former. Depending on the result in this step, remaining samples are plotted in the third and final step on either of Figures 3.9c or 3.10b, which separate initial from late oceanic arc potassic igneous rocks based essentially on the lower La content of the former; or on Figures 3.9d or 3.10c, which separate continental from postcollisional potassic igneous rocks based on slightly lower Ce/P ratios in the latter. However, since both the final settings involve destruction of oceanic crust in a subduction setting, they not unexpectedly generate potassic igneous rocks of similar composition, and therefore overlaps are unavoidable.

3.7 Theoretical Basis for Discrimination Between Potassic Igneous Rocks in Different Tectonic Settings

This discussion attempts to provide a theoretical foundation for the empirical observations in previous sections. It examines those factors which can, in principle, be expected to induce differences between potassic igneous rocks erupted in different tectonic settings, and shows how the geochemical differences outlined above can be derived. As genetic processes in arc and within-plate settings are so distinct, arc potassic igneous rocks are discussed first.

Despite the amount of literature published about high-K rocks, the petrogenetic processes producing the various types of potassic igneous rocks are still debated (e.g. Peccerillo 1992). However, there is general consensus that potassic magmas cannot be derived by partial melting of normal mantle peridotite, but require heterogeneous mantle sources which have been metasomatically enriched in LILE and LREE (Edgar 1987; Foley and Peccerillo 1992).

Potassic igneous rocks are commonly enriched in LILE, LREE, volatiles, and halogens, particularly Cl and F (Aoki et al. 1981; Foley 1992; Müller and Groves 1993). These elements are mainly incorporated in the hydrous phenocrysts such as phlogopite and/or amphibole (Aoki et al. 1981). As a result, portions of the Earth's mantle rich in phlogopite ± clinopyroxene are considered to be important in the genesis of potassic melts (Edgar and Arima 1985; Foley 1992). Meen (1987) considers that potassic igneous rocks form by low degrees of partial melting, under hydrous conditions in a low heat-flow environment, of upper mantle lherzolite that has been metasomatically enriched in such elements as LILE and LREE. In arc settings, the partial melting may be achieved by modification of mantle geotherms in proximity to the cold subducted slab (Taylor et al. 1994), while the metasomatic enrichment may be achieved by overprinting and veining of the mantle wedge by either volatile- and LILE-enriched *fluids* and/or by actual alkalic, low-temperature, LILE- and LREE-enriched partial *melts* derived during dehydration of the subducted oceanic slab

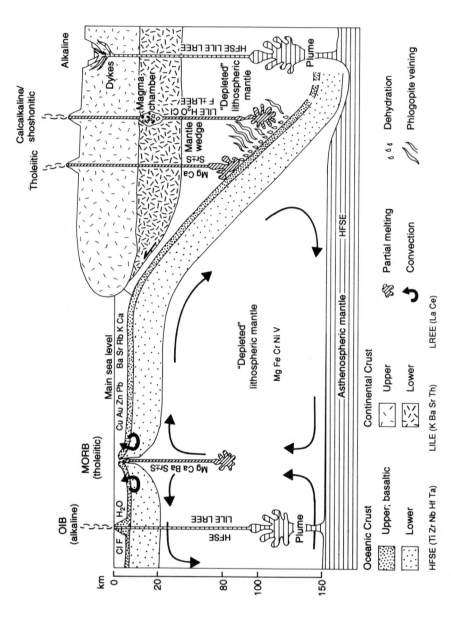

Fig. 3.11.   Schematic cross section showing the main components of a continental arc and distribution of elements (see also Müller et al. 1992b).

(Saunders et al. 1980; Pearce 1983; Bailey et al. 1989; Sun and McDonough 1989). Based on studies of mantle xenoliths (Bailey 1982; Menzies and Hawkesworth 1987), the metasomatically introduced volatiles, LILE, and LREE are preferentially sited in hydrous minerals such as phlogopite, amphibole, and apatite which are concentrated either in veins or dispersed throughout the upper mantle peridotite (Peccerillo 1992; Ionov and Hofmann 1995). These minerals have lower melting temperatures than the surrounding mantle rocks and partial melting may preferentially affect these metasomatic veins, thus generating potassic magmas.

Once melting has been initiated, the interplay of two further factors is then generally considered to influence the chemistry of *all* subduction-related melts actually produced:

- Differing rates and/or angles of the subduction process (Saunders et al. 1980; Rock et al. 1982).
- Differing relative inputs from at least three identified source components (Wheller et al. 1986):
    1. *Subducted oceanic crust*, characterized by high LILE/LREE (e.g. high Sr/Nd and Rb/Nd) and high LILE/HFSE ratios (e.g. high Ba/Nb and Th/Ta).
    2. *Subducted marine sediment*, characterized by low Sr/Nd (~ 9) and high Th/Ta ratios (> 100) (Rogers et al. 1985), high Pb, Ba, and La contents (Sun and McDonough 1989), and perhaps by negative Eu anomalies (McLennan and Taylor 1981).
    3. *The overlying mantle wedge*, characterized by low Rb (< 50 ppm) and especially Pb contents (Ellam and Hawkesworth 1988).

Melts may be further influenced by the nature of the crust through which they must pass before eruption, if any contamination or assimilation takes place. There is also the classic increase in K in arc-related suites with increasing distance from the subduction trench, the cause of which remains controversial. This may be accompanied by sympathetic increases in Th, Ta, and Nb (cf. Brown et al. 1984), attributed by some to the derivation of liquids from heterogeneous mantle which changes characteristics from subduction-related to within-plate type away from the trench. A schematic cross-section showing the distribution of the elements in a subduction zone (continental arc) is illustrated in Figure 3.11.

On this basis, it is possible to account for some of the observed differences among the four arc settings. As indicated in Figures 3.9 and 3.10, continental and postcollisional arc potassic igneous rocks are enriched in Zr, Hf, Nb, and LREE; they also have higher Sr and Ba contents, and higher K/Na ratios than oceanic arc potassic igneous rocks (cf. Fig. 3.6). This might be explained by a greater role for the alkalic *melt*-induced metasomatism of the mantle wedge, which is believed to yield stronger enrichments in these particular elements than *fluid*-induced metasomatism (Bailey et al. 1989). Such an augmented role might further explain the progressive transition of potassic igneous rocks to more strongly alkaline (i.e. undersaturated) magmatism in some postcollisional arc settings (see Chap. 4.2.). The passage of continental and postcollisional potassic igneous rocks through thick continental crust, as opposed to relatively thin depleted MORB in the case of oceanic arc potassic igneous rocks, could also account, in part, for their higher LILE and LREE contents. However, enrichments in LREE can also be explained by metasomatic fluids enriched in volatiles such as chloride, with chloride complexes preferentially transporting the LREE (Campbell et al. 1998).

Continental arc potassic igneous rocks have slightly higher Rb, Sr, Ba, and Ce, but lower Nb and P contents, than postcollisional arc potassic igneous rocks: see positive Sr anomalies of the former on Figure 3.5a. These differences are not easy to explain, but may result from the above-mentioned progressive transition to more alkaline magmatism in some postcollisional arcs.

Oceanic arc potassic igneous rocks generally have the lowest concentrations of LILE (i.e. < 310 ppm Rb, < 1500 ppm Sr, < 1500 ppm Ba), LREE (e.g. < 115 ppm La, < 150 ppm Ce), and HFSE (e.g. < 300 ppm Zr, < 20 ppm Nb, < 5 ppm Hf). This may be explained via a mainly *fluid*-derived metasomatic enrichment of the underlying mantle wedge in oceanic arc settings; this does not significantly increase LREE and HFSE concentrations of mantle material because these elements are not mobilized due to their retention in insoluble phases in the subducted plate (Briqueu et al. 1984; Bailey et al. 1989). The relatively low LILE concentrations of the rocks may reflect their origin in an environment where oceanic crust, derived from a depleted mantle, has been subducted beneath the oceanic crust of another plate. During uprise, the melts produced in this setting must pass through a relatively thin oceanic crust consisting of depleted MORB. This could explain their very low concentrations of LILE compared to magmas from within-plate settings, where the melts have to pass through a relatively thick continental crust during their uprise. The probability of crustal assimilation for within-plate related potassic igneous rocks, and a resulting enrichment in LILE, is therefore much higher.

For within-plate settings, the partial melting required to generate potassic igneous rocks may be caused by processes such as pressure-release during intraplate rifting (Nelson et al. 1986; Leat et al. 1988), or by asthenospheric upwelling associated with lithospheric thinning. The required metasomatic enrichment of the source, in turn, may be induced by the local invasion of regions within the subcontinental mantle by incompatible-element-enriched mantle-plumes. The plumes are probably derived from sources near the 650-km seismic discontinuity or near the core-mantle boundary (Ringwood 1990), and have heads that can be as large as 400 km^2 (Halliday et al. 1990). Alternatively, the enrichment may reflect pre-existing, long-term, intrinsic heterogeneities within the upper mantle and bear no relation to the actual melting event. At any rate, small degrees of partial melting of phlogopite-bearing mantle peridotite, at depths below the level of amphibole stability and in the presence of CO_2, are believed to generate potassic melts (Nelson et al. 1986). The fact that within-plate potassic igneous rocks show the highest HFSE contents (up to 5.5 wt % TiO_2, 840 ppm Zr, 74 ppm Nb, 30 ppm Hf) of the investigated settings and very high Nb/Y ratios (up to 3) may reflect this assumed greater role for CO_2, which mobilizes many HFSE (Pearce and Norry 1979). Within-plate potassic igneous rocks are also characterized by the highest LILE (i.e. up to 640 ppm Rb, 4600 ppm Sr, 9000 ppm Ba) and LREE contents (e.g. up to 230 ppm La, 390 ppm Ce) of all investigated settings, but geochemical explanations for this are obscure.

In areas where the composition of potassic igneous rocks results from mixing processes between melts from both asthenospheric (characterized by within-plate geochemistry) and lithospheric (subduction-modified) mantle sources (e.g. Wyoming Province; see McDonald et al. 1992; Gibson et al. 1993), the geochemical discrimination of the tectonic setting becomes problematical. In those areas, some asthenosphere-derived magma plumes might have been contaminated by pockets of subduction-metasomatized lithosphere during uprise, whereas others remained relatively uncontaminated (Gibson et al. 1993).

3.8   Conclusions

- Young potassic igneous rocks, from the five main tectonic settings in which they occur, have somewhat different major- and trace-element compositions.
- Potassic igneous rocks from within-plate settings — such as the western USA — are very distinctive, due to very high concentrations of LILE (e.g. Rb, Sr, Ba), LREE (e.g. La, Ce, Sm) and HFSE (notably Ti, Zr, Nb, Hf).
- Potassic igneous rocks from oceanic arc settings have the lowest concentrations of LILE, LREE and HFSE of all investigated settings, and those from initial and late oceanic arc settings can be discriminated by the higher P, Zr, Nb, and La concentrations of the latter.
- Potassic igneous rocks from continental and postcollisional settings show the most subtle differences, but can still be distinguished by the slightly higher Sr, Zr, and Ce concentrations of the former.
- The above-mentioned differences extend to immobile elements, so it is possible, in principle, to identify the tectonic setting of older potassic igneous rocks from their geochemistry, even where direct geological evidence is equivocal. This is best done in a hierarchical set of diagrams, or by formal stepwise discriminant analysis, which successively strips off the most distinctive compositions in order to progressively discriminate more subtle differences.
- Potassic igneous rocks in certain continental, postcollisional, and late oceanic arc settings are associated with world-class deposits of gold and/or base metals (Chaps 6, 7), and hence these new discrimination methods may be useful in exploration as well as in tectonic and petrogenetic studies. By identifying those older potassic igneous rocks which have the most favourable tectonic settings, it may be possible to identify, more efficiently, those terranes which have economic potential for specific styles of mineralization associated with high-K igneous rocks. This is discussed further in Chapter 6.

4   Selected Type-Localities of Potassic Igneous Rocks from the Five Tectonic Settings

4.1   Roman Province (Italy): Example from a Continental Arc Setting

4.1.1   Introduction

The volcanic rocks of Central Italy (Fig. 4.1) are divided into three magmatic provinces (van Bergen et al. 1983):
- In the north, predominantly acid igneous rocks of Amiata, Roccastrada, San Vincenzo, and Elba represent the *Tuscan* Province.
- In Central Italy, along the western side of the Apennine Fold Belt, the highly potassic volcanic centres of Vulsini, Vico, Sabatini and Alban Hills form the *Roman* Province (see also Holm et al. 1982; Rogers et al. 1985).

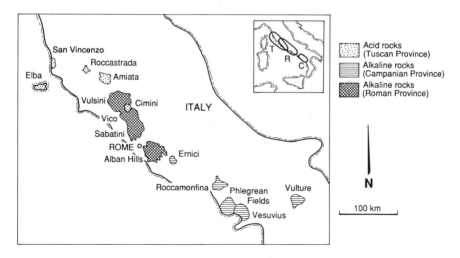

Fig. 4.1.   Geological overview of the Roman Province, Central Italy. Provinces: T = Tuscan; R = Roman; C = Campanian. Modified after Holm et al. (1982) and van Bergen et al. (1983).

Table 4.1. Representative whole-rock major- and trace-element geochemistry of potassic igneous rocks from the Roman Province, Central Italy. Major elements are in wt %, and trace elements are in ppm. Fe_2O_3 (tot) = total iron calculated as ferric oxide. Data from Civetta et al. (1981), Holm et al. (1982), Rogers et al. (1985), and Conticelli and Peccerillo (1992).

Province/deposit:	Vulsini, Roman	Vulsini, Roman	Sabatini, Roman	Mount Ernici, Roman	Mount Ernici, Roman
Location:	Italy	Italy	Italy	Italy	Italy
Rock type:	Leucitite	Leucitite	Tephrite	Basalt	Leucitite
Tectonic setting:	Continental arc	Continental arc	Continental arc	Continental arc	Continental arc
Reference:	Holm et al. (1982)	Rogers et al. (1985)	Conticelli & Peccerillo (1992)	Civetta et al. (1981)	Civetta et al. (1981)
SiO_2	47.16	47.52	48.36	48.52	47.39
TiO_2	0.74	0.87	0.70	0.77	0.72
Al_2O_3	12.46	14.52	16.80	16.35	17.85
Fe_2O_3 (tot)	7.49	8.44	6.95	6.87	6.03
MnO	0.11	0.14	0.13	0.15	0.13
MgO	9.22	7.38	6.57	9.03	6.36
CaO	15.41	13.15	9.85	12.03	10.53
Na_2O	0.68	0.96	1.30	2.79	2.51
K_2O	4.67	5.14	8.33	2.60	7.36
P_2O_5	0.22	0.33	0.61	0.27	0.54
LOI	1.46	1.45	0.40	0.62	0.57
Total	99.62	99.90	100.00	100.00	99.99
V	225	243	n.a.	233	233
Cr	295	138	316	490	151
Ni	116	61	74	87	58
Rb	356	425	636	112	335
Sr	784	1122	1812	848	1412
Y	23	26	26	n.a.	n.a.
Zr	187	180	266	86	218
Nb	11	6	14	8	9
Ba	853	592	1202	500	892
La	62	56	88	n.a.	n.a.
Ce	112	127	202	n.a.	n.a.
Th	25	22	46	10	28
Ta	n.a.	0.6	0.6	n.a.	n.a.
Hf	n.a.	5	6	3	6

– Potassic volcanoes such as Roccamonfina and Vesuvius, and the lavas of the Phlegrean Fields and Vulture form the *Campanian* Province in the south (van Bergen et al. 1983; D'Antonio and Di Girolamo 1994).

The Pliocene-Quaternary volcanic rocks from the Roman Province, Italy, are considered to be typical examples of orogenic ultrapotassic rocks as defined by Foley et al. (1987).

4.1.2 Regional Geology

Despite the potassic igneous rocks of the Roman Province having been studied by many petrologists (e.g. Appleton 1972; Cundari 1979; Civetta et al. 1981; Holm et al. 1982; van Bergen et al. 1983; Di Girolamo 1984; Poli et al. 1984), their petrogenesis has been the subject of controversy for many years (Rogers et al. 1985). There has also been considerable debate concerning the tectonic setting of these rocks. Most workers have interpreted the potassic igneous rocks to be related to subduction beneath the Calabrian Arc (Ninkovich and Hays 1972; Edgar 1980), but this has been contested by Cundari (1979). Modern studies based on stable isotopes (Rogers et al. 1985) seem to confirm the importance of subduction processes in the genesis of these rocks.

The Vulsinian District of the Roman Province is by far the largest in the region, covering 2280 km^2 (von Pichler 1970), and is the most intensively studied (Holm et al. 1982; Rogers et al. 1985). The volcanic activity is represented by lava flows and pyroclastic rocks such as tuffs and ignimbrites (Holm et al. 1982).

4.1.3 Mineralogy and Petrography of the Potassic Igneous Rocks

The potassic igneous rocks from the Roman Province consist mainly of latites, tephrites, trachytes, phonolites, and leucitites (Holm et al. 1982).

Most of the lavas have porphyritic textures with phenocrysts of clinopyroxene, plagioclase, and leucite with minor sanidine and olivine in a fine-grained groundmass consisting of plagioclase, leucite, and clinopyroxene (Holm et al. 1982). The more silica-saturated latites and trachytes are characterized by similar assemblages, although leucite is absent and sanidine and quartz present; they also have biotite and apatite phenocrysts (Holm et al. 1982; Rogers et al. 1985).

4.1.4 Geochemistry of the Potassic Igneous Rocks

Most of the lavas from the Roman Province are highly potassic (Conticelli and Peccerillo 1992), and some can be defined as ultrapotassic with K_2O and MgO > 3 wt %, and K_2O/Na_2O ratios > 2 (Foley et al. 1987). The compositions (Table 4.1) normally vary from silica-undersaturated to moderate SiO_2 contents (47– 56 wt %); TiO_2 contents are low (< 0.8 wt %) and Al_2O_3 contents are variable (12–

17 wt %), but can be as high as 19.9 wt % (e.g. Rogers et al. 1985), which is typical for subduction-related potassic igneous rocks (Morrison 1980).

The rocks have high concentrations of LILE (e.g. up to 636 ppm Rb, up to 1812 ppm Sr), intermediate LREE (e.g. < 100 ppm La, < 200 ppm Ce), and low HFSE (< 0.8 wt % TiO_2, < 14 ppm Nb, < 6 ppm Hf; see Table 4.1), when compared to those potassic igneous rocks derived from within-plate settings (Müller et al. 1992b). Based on their geochemistry, the rocks are interpreted to be subduction-related and they occur in a continental-arc setting, as previously suggested by most petrologists concerned with the area (e.g. Edgar 1980). This interpretation is not consistent with studies by Cundari (1979), who proposes a within-plate setting for the rocks from the Roman Province. However, potassic igneous rocks from within-plate settings are typically characterized by very high HFSE concentrations (cf. Müller et al. 1993a), which is not the case in the rocks from the Roman Province (see Table 4.1).

4.2   Kreuzeck Mountains, Eastern Alps (Austria): Example from a Postcollisional Arc Setting

4.2.1   Introduction

This region is described in more detail because data for the potassic igneous rocks were collected specifically for this study (cf. Müller 1993), the tectonic setting of the rocks is complex, and their precious-metal contents are discussed in Chapter 5.

The major components of the Eastern Alps in Austria — and adjoining areas of Switzerland, Italy, and Yugoslavia — are the Northern and Southern Calcareous Alps and the Central Alps (Fig. 4.2).

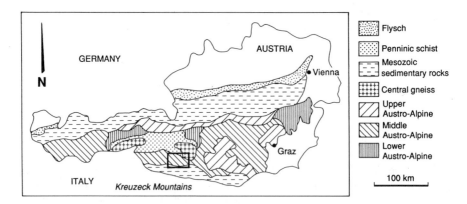

Fig. 4.2.   Geological overview of the Eastern Alps, Austria, showing the location of the Kreuzeck Mountains. Modified after Reimann and Stumpfl (1985).

European-African plate collision and Alpine nappe emplacement in the late Creta-
ceous-Eocene were followed by a significant phase of mafic to felsic Oligocene
dyke magmatism, mostly concentrated along a major (700 km), east-west-trending
Tertiary suture, the Periadriatic Lineament (Exner 1976). This lineament, also known
as the Insubric Line, forms a dextrally transpressive intracontinental branch of the
Europe-Africa plate boundary (Laubscher 1988).

Dykes cut Austro-Alpine, South Alpine, and, rarely, Penninic units. They range in
composition from basaltic to rhyolitic, but are mainly calc-alkaline andesitic and
basaltic. Their compositions change across broad zones, from tholeiitic and calc-
alkaline in the Southeastern Alps to high-K calc-alkaline in the Central Alps and to
shoshonitic and ultrapotassic in the northwestern and western sector (Beccaluva et
al. 1983).

4.2.2 Regional Geology

The Kreuzeck Mountains, southern Austria, are composed of rocks of the middle
Austro-Alpine unit of the Central Alps (Fig. 4.2) and consist of polymetamorphic
crystalline basement rocks, which are partly overlain by a lower Palaeozoic
volcanosedimentary sequence metamorphosed to greenschist facies (Reimann and
Stumpfl 1981, 1985). The area studied covers more than 600 km^2 between the Möll
Valley in the north and the Drau Valley in the south, and east-west from Iselsberg to
Möllbrücke (Fig. 4.3).

The Kreuzeck Mountains are situated in the suture zone between the African and
the European plates (Hoke 1990). Mica-schists, gneisses, and amphibolites are the
main rock types; no quantitative age data are available, but the protoliths are estimated to be older than Permo-Carboniferous (Lahusen 1972). In the prevailing plate
tectonic models, this Austro-Alpine block forms part of the Adriatic plate which
overrode South Penninic units during the Eocene continent-continent collision. These
Penninic units, with oceanic metasedimentary rocks, are exposed in the Tauern Window to the north. The Oligocene is characterized by several generations of dykes
and local granodioritic plutons during a phase of extensional tectonics. This inhomo-
geneous extensional regime is followed by a Miocene compression (Laubscher 1988).

The rocks of the Kreuzeck Mountains reveal a homogeneous pattern of deformation. Only in the northeastern area, which is nearest to the southern edge of the
Tauern Window, are different phases of deformation detectable (Oxburgh 1966).
Predominant tectonic features are faults and fractures, which show a dominant trend
striking east-west. This correlates with the strike direction of most felsic dykes and
some of the mafic dykes investigated here, which follow zones of weakness or fracture zones in the host rock, and implies a close connection with the east-west-strik-
ing Periadriatic Lineament.

The entire period of Oligocene orogenic magmatism is linked to continent-continent collision of the African and the European plates after the subduction of the
Penninic oceans (Deutsch 1986). This includes the Western Alps (Venturelli et al.
1984), and dyke swarms along the southern margin of the European plate, which are

believed to be related to northwesterly dipping subduction of African oceanic lithosphere (Beccaluva et al. 1983). This event also produced back-arc spreading in the southwestern area (Provence, Balearic Basin, Sardinia).

More than 60 former prospects and mines are known from the Kreuzeck area (Friedrich 1963); mining activities date back to the Middle Ages and were mainly directed at stratabound ores of antimony-tungsten, mercury, and copper-silver-gold (Reimann and Stumpfl 1981). Most ore deposits are in the southern part of the Kreuzeck Mountains within a sequence of metavolcanic and metasedimentary rocks (Lahusen 1972) which extends, with tectonic interruptions, for over 40 km along the Drau Valley (Fig. 4.3).

Early work by Friedrich (1963) suggested a relationship between ore deposits and Tertiary felsic porphyritic dykes in the Kreuzeck Mountains. He believed the mineralization to be of epigenetic hydrothermal origin, related to a hypothetical Tertiary pluton underlying the area. He interpreted numerous small granodioritic intrusions,

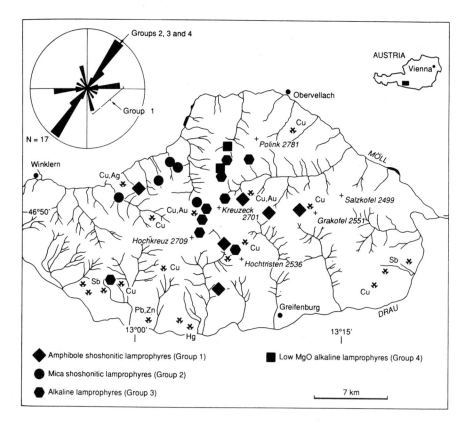

Fig. 4.3.   Geographic map of the Kreuzeck Mountains, Austria, showing major mining areas and the location of shoshonitic and alkaline lamprophyres. Modified after Müller et al. (1992a).

as well as the lamprophyres and felsic dykes, as an indication of the presence of this postulated pluton. Later investigations by Höll and Maucher (1968) interpreted most of the antimony deposits to be submarine-exhalative synsedimentary in origin (lower Palaeozoic age) and partly remobilized during Variscan and Alpine metamorphism, an opinion supported by Lahusen (1972).

4.2.3 Mineralogy and Petrography of the Lamprophyres

Dyke rocks of the northern and central Kreuzeck Mountains are dominated by mafic types (lamprophyres, basaltic dykes), whereas in the southern part, most dykes are felsic (microdioritic). The lamprophyres of the Kreuzeck Mountains are unfoliated dykes which normally cut their host rocks discordantly. The dykes mainly strike northeast-southwest, although some strike east-west, parallel to the Periadriatic Line, which is about 15 km south of the Kreuzeck Mountains (Müller et al. 1992a). The thickness of the mafic dykes normally varies from 0.5 to 5 m, although dykes of intermediate composition can be up to 10 m thick.

Three petrographic types, based on phenocryst mineralogy, can be recognized (cf. Müller et al. 1992a):

- Type 1, amphibole-clinopyroxene±mica phyric.
- Type 2, mica-clinopyroxene±amphibole phyric.
- Type 3, mica-olivine±clinopyroxene±amphibole phyric.

Mica and amphibole phenocrysts are generally about 0.5–4 mm long, and commonly show compositional zonation (Fig. 4.4). The lamprophyres are characterized by a fine-grained microcrystalline groundmass comprising plagioclase, clinopyroxene, amphibole, and mica, with less common alkali-feldspar and apatite. Plagioclase is typically saussuritized and the mafic minerals are commonly altered to chlorite. Partially resorbed quartz xenocrysts, presumably derived from basement rocks, are present in some samples, particularly those of petrographic type 1.

According to the classification scheme of Mogessie et al. (1990), amphiboles from rocks of petrographic type 1 are tschermakitic hornblendes and tschermakites with 0.8–2.4 wt % TiO_2 (Table 4.2). According to Rock (1991), tschermakitic amphiboles are characteristic of calc-alkaline lamprophyres whereas Ti-rich amphiboles, especially kaersutite, are diagnostic of alkaline lamprophyres. Amphibole mineral chemistry thus indicates that dykes of both alkaline and calc-alkaline affinity are present in the Kreuzeck area.

Micas analysed from rocks of petrographic types 2 and 3 are phlogopites with mg# >75 (Table 4.2). They have high TiO_2 contents (~ 4–6 wt %), which are more typical of those for alkaline lamprophyres (see Rock 1991; Fig. 4.5).

Table 4.2.   Microprobe (energy-dispersive spectra - EDS) analyses of selected phenocrysts from lamprophyres and basaltic dykes from the Kreuzeck Mountains, Eastern Alps, Austria. Oxides and Cl are in wt %. FeO (tot) = total iron calculated as ferrous oxide, Hbl. = Hornblende, Tsch. = tschermakitic, Tscherm. = tschermakite, Ox. form. = oxygen formula. Sample numbers refer to specimens held in the Museum of the Department of Geology and Geophysics, The University of Western Australia. From Müller et al. (1992a).

Sample no.:	119036	119037	119049	119050	119051[a]	119051[b]
Petrographic type:	3	1	1	1	1	1
Geochemical group:	3	1	1	1	1	1
wt %						
SiO_2	39.35	43.85	43.52	41.85	42.42	41.94
TiO_2	4.63	2.47	0.79	1.81	1.77	2.20
Al_2O_3	13.10	11.25	14.04	13.91	13.59	10.53
Cr_2O_3	0.10	0.43	0.10	0.09	0.21	0.10
FeO (tot)	12.18	8.40	8.12	15.98	8.88	18.68
MnO	0.12	0.10	0.11	0.32	0.11	0.42
MgO	11.60	16.33	16.03	10.63	15.66	10.26
CaO	11.82	11.47	11.59	10.24	11.92	10.45
Na_2O	1.98	2.20	1.58	1.62	2.01	2.02
K_2O	1.55	0.36	0.49	0.70	0.89	0.68
Cl	0.05	0.04	0.05	0.05	0.04	0.04
Total	96.48	96.90	96.42	97.20	97.50	97.32
Name	Kaersutite	Tsch. Hbl	Tscherm.	Tscherm.	Tscherm.	Tscherm.
mg#	62.9	77.5	77.9	54.2	75.8	50.5
Ox. form.	23					
Atoms						
Si	5.949	6.276	6.335	6.272	6.061	6.397
Ti	0.527	0.267	0.087	0.204	0.189	0.253
Al	2.334	1.899	2.409	2.457	2.292	1.892
Cr	0.009	0.052	-	-	0.026	0.012
Fe	1.544	1.006	0.988	2.003	1.064	2.383
Mn	0.018	0.017	-	0.041	0.017	0.054
Mg	2.615	3.473	3.477	2.374	3.339	2.332
Ca	1.916	1.762	1.808	1.643	1.829	1.708
Na	0.581	0.611	0.444	0.472	0.558	0.596
K	0.301	0.069	0.091	0.131	0.163	0.132
Cl	-	-	-	-	-	-
Total	15.794	15.432	15.641	15.597	15.538	15.761

[a] Core.   [b] Rim.

4.2.4 Geochemistry of the Lamprophyres

Major- and trace-element chemistry is listed in Table 4.3. The dykes have a range of SiO_2 contents (42.6–57 wt %), and on a total alkalis versus silica plot (Fig. 4.5) they

119055	119059	119061	119036	119042	119054	119056
3	3	1	3	3	3	2
3	2	1	3	3	3	4
40.95	45.19	43.48	38.21	36.64	38.22	37.13
2.80	2.21	1.78	4.05	6.18	5.24	6.11
14.95	10.14	13.41	15.94	16.25	15.91	15.57
0.09	0.10	0.10	1.09	0.10	0.23	0.09
14.71	8.59	7.81	6.85	9.94	8.56	10.41
0.26	0.11	0.11	0.11	0.12	0.10	0.10
8.67	16.85	15.98	20.05	17.42	19.34	17.89
10.17	11.05	11.82	0.08	0.09	0.08	0.08
2.28	2.01	1.91	0.38	0.82	0.70	0.42
2.26	1.04	1.25	9.52	8.79	8.78	9.25
0.04	0.05	0.05	0.05	0.05	0.04	0.04
97.18	97.34	97.70	96.33	96.40	97.20	97.09
Pargasite	Tsch. Hbl	Tscherm.	Phlogopite	Phlogopite	Phlogopite	Phlogopite
51.2	77.7	78.5	83.4	75.7	80.1	75.4
			22			
6.173	6.562	6.289	5.457	5.304	5.426	5.351
0.317	0.241	0.194	0.434	0.673	0.559	0.662
2.652	1.736	2.285	2.684	2.772	2.661	2.644
0.009	-	-	0.124	-	0.026	-
1.856	1.044	0.945	0.818	1.204	1.016	1.255
0.036	-	-	-	-	-	-
1.946	3.648	3.445	4.268	3.758	4.091	3.842
1.639	1.718	1.832	-	-	-	-
0.669	0.565	0.535	0.107	0.231	0.192	0.118
0.435	0.193	0.231	1.735	1.623	1.591	1.699
-	-	-	-	-	-	-
15.732	15.709	15.756	15.627	15.565	15.562	15.571

cluster into a mostly *nepheline*-normative alkaline group (< 48 wt % silica, 4–7 wt % alkalis) and a *hypersthene*-normative calc-alkaline group (> 48 wt % silica, 3–5.5 wt % alkalis). On the basis of their major- and trace-element chemistry and petrographic character, the rocks can be further subdivided into geochemical groups as described below (cf. Müller et al. 1992a).

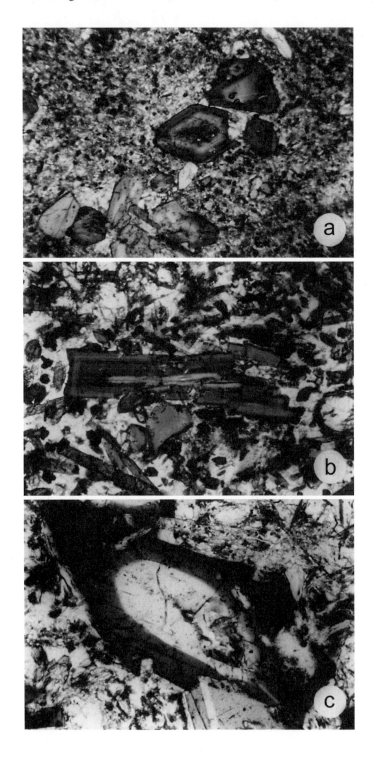

Fig. 4.4.   Photomicrographs (crossed nicols) of typical lamprophyre samples from the Kreuzeck Mountains, Austria. The field of view (FOV) is given in square brackets. (a) Porphyritic texture of tschermakite-bearing shoshonitic lamprophyre (119050) [FOV 3.5 mm]; the tschermakite phenocrysts show multiple zoning with Fe-rich rims and Mg-rich cores. (b) Porphyritic texture of phlogopite-bearing alkaline lamprophyre (119040) [FOV 2 mm]; the phlogopite phenocryst is zoned with dark brown Fe-rich rim and pale Mg-rich core. (c) Zoned tschermakite with Fe-rich rim and Mg-rich core (119051) [FOV 2 mm]. Sample numbers refer to specimens held in the Museum of the Department of Geology and Geophysics, The University of Western Australia.

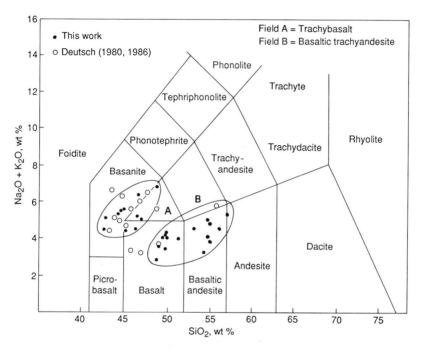

Fig. 4.5.   ($Na_2O + K_2O$) versus SiO_2 (TAS) plot of dykes in the Kreuzeck Mountains, Austria, from this study and Deutsch (1980, 1986). The dykes are clearly divided into two groups: a SiO_2-rich calc-alkaline to shoshonitic group and a less SiO_2-rich alkaline group. Modified from Müller et al. (1992a).

Group 1, amphibole-bearing shoshonitic lamprophyres (petrographic type 1). These rocks have variable K_2O (~ 0.8–2.6 wt %)and Rb (36–200 ppm) contents, suggesting some alkali loss. This is consistent with the generally altered nature of the groundmass in these dykes; that is, groundmass phlogopite is completely altered to chlorite in all samples except one, and feldspars are partially saussuritized in all samples. The low F content (< 10 ppm) of one sample is also consistent with alteration of mica to chlorite. MgO contents range from 3.6 to 9.8 wt %, mg# is ~ 50–70,

Table 4.3.   Representative whole-rock major- and trace-element geochemistry of investigated lamprophyres from the Kreuzeck Mountains, Austria. Major elements are in wt %, trace elements are in ppm, and precious metals are in ppb. Fe_2O_3 (tot) = total iron calculated as ferric oxide, n.a. = not analyzed, ne% = nepheline-normative content in wt % (calculated by CIPW-Norm) based on whole-rock geochemistry. Sample numbers refer to specimens held in the Museum of the Department of Geology and Geophysics, The University of Western Australia. From Müller et al. (1992a).

Sample no.:	119049	119045	119052	119053
Geochemical group:	1	2	3	4
SiO_2	55.91	54.53	46.50	46.40
TiO_2	0.45	1.38	2.01	2.41
Al_2O_3	19.83	14.56	14.14	15.33
Fe_2O_3 (tot)	6.17	7.21	9.54	11.73
MnO	0.13	0.08	0.15	0.20
MgO	3.94	5.67	8.20	5.69
CaO	6.04	4.43	8.32	8.06
Na_2O	1.97	4.30	2.33	2.55
K_2O	2.56	1.76	2.93	3.84
P_2O_5	0.11	0.24	0.53	0.45
LOI	3.13	5.89	5.46	3.30
Total	100.18	99.98	100.10	99.91
ne%	0.00	0.00	0.00	3.40
mg#	59.8	64.7	66.7	53.1
F	n.a.	210	150	520
Sc	9	7	15	18
V	148	183	303	347
Cr	49	272	326	133
Ni	5	121	186	20
Cu	7	11	18	14
Zn	62	74	100	97
As	< 2	2	3	< 2
Rb	208	40	87	70
Sr	286	690	794	998
Y	17	19	37	29
Zr	92	183	259	215
Nb	8	16	33	17
Sb	0.4	11.0	0.7	0.3
Ba	284	2821	1212	1327
W	120	130	81.0	100.0
Pb	23	32	25	25
Th	29	11	25	16
Pd	< 1	3	3	< 1
Pt	< 5	< 5	< 5	< 5
Au	< 3	26.8	15.5	< 3

and Ni contents vary from < 5 to 116 ppm, indicating that Group 1 dykes include both evolved and relatively primitive compositions. Group 1 dykes have 0.5–1.0 wt % TiO_2 and < 13 ppm Nb. Barium contents are < 400 ppm, resulting in the low Ba/Rb (< 10) and Ba/Nb (< 43) ratios characteristic of subduction-derived potassic magmas.

Group 2, mica-bearing shoshonitic lamprophyres (petrographic types 2 and 3). As with Group 1, these rocks have somewhat variable K_2O contents (~ 0.8–2.2 wt %) and K_2O/Na_2O ratios (~ 0.2–1.2) mainly reflecting alteration of groundmass mica and alkali feldspar. Whole-rock F contents (< 500 ppm) are low for typical mica-lamprophyres and suggest F loss. Olivine-phyric samples have high MgO contents (> 10 wt %), mg# (> 70), and Ni contents (> 400 ppm), indicating that they represent primitive magma compositions. Other Group 2 rocks are more evolved — one sample has a mg# of ~ 65 and Ni contents of ~ 120 ppm. TiO_2 and Nb abundances are 1.0–1.5 wt % and < 17 ppm, respectively. The high Ba contents result in high Ba/Rb (> 20) and Ba/Nb (> 100) ratios, which clearly distinguish them from Group 1 rocks; such ratios are typical of mica-phyric shoshonitic magmas generated in subduction zone settings (e.g. Luhr et al. 1989).

Group 3, alkaline lamprophyres (petrographic type 3). The alkaline lamprophyres are mostly nepheline normative in composition, and they have high alkali contents with K_2O/Na_2O > 1. Mg# is consistently ~ 62–67 and the Ni content is ~ 150 ppm, suggesting that magma compositions are only slightly evolved. The HFSE contents are high (1.5–2.1 wt % TiO_2, > 200 ppm Zr, 30–55 ppm Nb), which clearly distinguishes these rocks from those of Groups 1 and 2. K/Nb ratios are ~ 500 and Ba/Nb ratios are ~ 35, higher than those present in most non-Dupal-type oceanic island basalt (OIB) sources which have K/Nb ratios of ~ 180 and Ba/Nb ratios of ~ 6 (Weaver et al. 1987; Sun and McDonough 1989). The trace-element ratios of Group 3 alkaline lamprophyres are, however, similar to, or slightly more elevated than, potassic, Dupal-type (EM2-type) OIB sources; for example, Gough Island rocks which have K/Nb ratios of ~ 430 and Ba/Nb ratios of ~ 16. It is thought that Dupal-type OIB sources involve recycled sedimentary material, ancient continental lithosphere, or subduction-zone metasomatized lithosphere (Sun and McDonough 1989). An important difference between Dupal-type OIBs and the Group 3 lamprophyres is the much lower TiO_2 contents of the latter: < 2.1 wt % compared with > 2.5 wt % for Dupal-type OIB.

Group 4, low mg# alkaline lamprophyres (petrographic type 2). These lamprophyres are characterized by low mg# (53–56) and low Ni contents (~ 20 ppm), indicating that they represent evolved magmas. TiO_2 contents are ~ 2.5 wt % and Zr and Nb contents are ~ 200 ppm and ~ 20 ppm, respectively. Because Zr and Nb contents are lower than in Group 3 lamprophyres, Group 4 rocks cannot be derived from Group 3 by fractional crystallization processes, and must have evolved from a separate primary magma. Group 3 dykes are clearly alkaline since they have > 5 wt %

Table 4.4.   Whole-rock K-Ar dating on lamprophyres and basaltic dykes from the Kreuzeck Mountains, Austria. The Rb-Sr data are from Deutsch (1984). The error on ages is quoted as ± 2σ. From Müller et al. (1992a).

Geochemical group	Sample no.	K-Ar age (Ma)	Rb-Sr age (Ma)	Initial $^{87}Sr/^{86}Sr$ (whole rock)	Orientation
1	119050	36 ± 1	-	-	E-W
2	119058	27 ± 0.5	31 ± 2	0.7079	NE-SW
3	119042	32 ± 0.7	29 ± 4	0.7055	NE-SW
4	119053	29 ± 0.7	-	-	NE-SW

total alkalis and > 2 wt % normative-nepheline component, yet their low Nb contents are typical of magmas derived from subduction zones. K/Nb (> 1300) and Ba/Nb (> 55) ratios are extreme for mafic alkaline magmas but similar to those of some highly potassic arc volcanoes such as Batu Tara in the Sunda Arc of Indonesia (Stolz et al. 1988; Foley and Wheller 1990). The TiO_2 contents of Group 4 lamprophyres are, however, significantly higher than those of subduction-zone derived potassic magmas, which typically have < 1.5 wt % TiO_2 .

A plot of Ba/Rb versus TiO_2 (Fig. 4.6) is effective in separating the dyke rocks into Groups 1 to 4.

K-Ar ages (Müller et al. 1992a) are between ca. 27 and 36 Ma (Table 4.4). The older shoshonitic lamprophyres (ca. 36 Ma) imply subduction-related features with low HFSE contents, whereas the younger alkaline lamprophyres (ca. 30 Ma) show

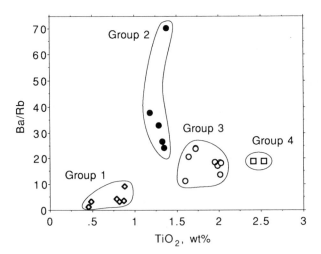

Fig. 4.6.   Ba/Rb versus TiO_2 plot showing the four geochemical groups of investigated dykes. From Müller et al. (1992a).

transitional characteristics with higher HFSE contents (see Table 4.3). The older age for a Group 1 lamprophyre is also consistent with the greater degree of alteration and evidence for recrystallization in the Group 1 samples, suggesting emplacement during the latter stages of Alpine metamorphism. Their mainly east-west dyke orientations also distinguish them from the slightly younger Group 2 to 4 lamprophyres.

The largely Oligocene age of these lamprophyres is much younger than the supposed age of mineralization in the area (older than Permo-Carboniferous).

4.3 Northern Mariana Arc (West Pacific): Example from an Initial Oceanic Arc Setting

4.3.1 Introduction

Intra-oceanic island arcs such as the Mariana Arc are restricted to the circum-Pacific region (Stern 1979). The Mariana Arc is located about 2000 km north of the mainland of Papua New Guinea in the West Pacific (Fig. 4.7) and consists of 21 major volcanoes and seamounts (Garcia et al. 1979; Meijer and Reagan 1981; Stern et al. 1988), from Nishino Shima in the north to Guam in the south. The subduction trench is situated in the west of the island arc.

Following the geographic distribution of volcanic islands and seamounts, the Mariana Arc is divided into three provinces: the Northern Seamount Province, the Central Island Province, and the Southern Seamount Province (Stern 1979; Lin et al. 1989).

4.3.2 Regional Geology

The magmatic system of the Mariana Arc formed by the subduction of the Pacific plate beneath the Philippine Sea plate (Lin et al. 1989). Volcanoes in the northern Mariana Arc between Uracas (latitude 20° N) and Minami Iwo Jima (24° N) are still active yet entirely submarine (Stern et al. 1988). In contrast to the mainly low-K basalts and andesites of the subaerially exposed volcanoes in the Central Island Province (Meijer and Reagan 1981), the northern and southern parts are dominated by high-K igneous rocks which have been classified as shoshonites by Stern et al. (1988) and Bloomer et al. (1989). The shoshonites apparently represent the youngest volcanic products of arc evolution, and were previously interpreted to be products of an episode of back-arc rifting (Stern et al. 1988; Bloomer et al. 1989; Lin et al. 1989). However, recent studies have demonstrated that the region now defined by the shoshonitic rocks occupies a "pre-rift" position (Baker et al. 1996; Gribble et al. 1998). The shoshonites occur along the magmatic front, which is unusual for an intra-oceanic arc (De Long et al. 1975) because high-K calc-alkaline rocks and shoshonites normally occur farthest from the trench and above the highest parts of the Benioff Zone (Morrison 1980).

Andesites from Sarigan, an island in the Central Island Province, have been dated at 0.5 ± 0.2 Ma (Meijer and Reagan 1981).

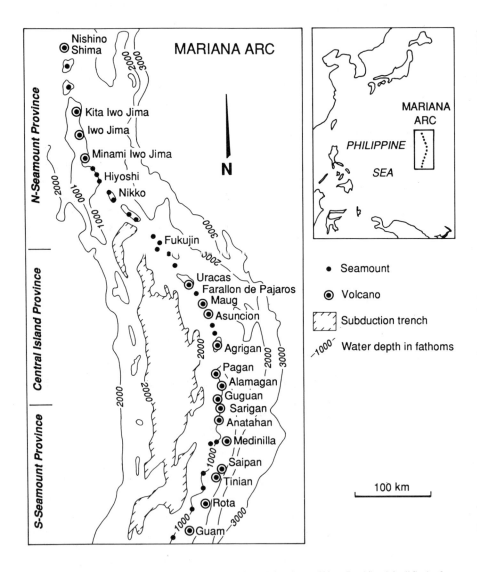

Fig. 4.7.   Geographic overview of the Mariana island arc, West Pacific. Modified after Stern (1979) and Bloomer et al. (1989).

4.3.3 Mineralogy and Petrography of the Potassic Igneous Rocks

The rocks are densely phyric and have highly porphyritic textures (Bloomer et al. 1989), and basalt and andesite dominate. The typical phenocryst assemblages are plagioclase, clinopyroxene, orthopyroxene, $\pm$ olivine, $\pm$ titanomagnetite, $\pm$ apatite (Bloomer et al. 1989; Gribble et al. 1998) in a groundmass comprising mainly plagioclase and clinopyroxene, and minor orthopyroxene and potassic feldspar (Meijer and Reagan 1981). Olivine is most common in the basalts of the Southern Seamount Province. Hornblende phenocrysts are rare and mainly restricted to the andesitic rocks (Meijer and Reagan 1981).

The shoshonitic basalts from the Northern and Southern Seamount Provinces contain phenocrysts of plagioclase, clinopyroxene, olivine, and biotite (Bloomer et al. 1989).

4.3.4 Geochemistry of the Potassic Igneous Rocks

The whole-rock geochemistry of the volcanic rocks from the Mariana Arc has been studied by Dixon and Batiza (1979), Stern et al. (1988), Bloomer et al. (1989), Lin et al. (1989), and Woodhead (1989).

The shoshonitic rocks, which are of interest here, are characterized by high K/Na ratios (> 0.6), high K_2O (up to 2.54 wt %), and relatively high Al_2O_3 contents (up to 18 wt %): see Table 4.5. They are further characterized by very low concentrations of LILE (e.g. commonly < 700 ppm Ba, < 34 ppm Rb), and very low LREE (e.g. < 40 ppm La, < 50 ppm Ce) and HFSE (e.g. < 90 ppm Zr, < 5 ppm Nb, < 2 ppm Hf) contents. These very low element abundances for the Mariana Arc shoshonites are distinct among the potassic igneous rock clan, and, in combination with their unusual generation during the initial stages of arc evolution (De Long et al. 1975; Stern et al. 1988), define those high-K rocks from initial oceanic arcs (Müller et al. 1992b).

4.4 Vanuatu (Southwest Pacific): Example from a Late Oceanic Arc Setting

4.4.1 Introduction

Vanuatu (formerly New Hebrides) is an isolated island arc about 2000 km east of Australia and forms part of the outer Melanesian Arc (Coleman 1970). The island arc consists of nine major islands and the subduction trench is situated to the west of the islands (Fig. 4.8). The islands are dominated by dome-shaped basaltic shield volcanoes (Coleman 1970).

Table 4.5.   Representative whole-rock major- and trace-element geochemistry of potassic igneous rocks from the Mariana island arc, West Pacific. Major elements are in wt %, and trace elements are in ppm. Fe_2O_3 (tot) = total iron calculated as ferric oxide. Data from Dixon and Batiza (1979), Bloomer et al. (1989), and Woodhead (1989).

Province/deposit:	Northern Seamount	Northern Seamount	Northern Seamount
Location:	Mariana Islands	Mariana Islands	Mariana Islands
Rock type:	Basalt	Basalt	Basalt
Tectonic setting:	Initial oceanic arc	Initial oceanic arc	Initial oceanic arc
Reference:	Dixon & Batiza (1979)	Bloomer et al. (1989)	Woodhead (1989)
SiO_2	53.37	50.25	54.70
TiO_2	0.73	0.80	0.86
Al_2O_3	16.01	18.54	16.78
Fe_2O_3 (tot)	9.21	8.54	11.03
MnO	0.16	0.21	0.24
MgO	4.90	4.23	3.29
CaO	9.93	9.31	8.07
Na_2O	3.00	2.97	3.34
K_2O	1.05	2.54	1.43
P_2O_5	0.19	0.34	0.27
LOI	0.35	1.80	n.a.
Total	98.90	99.53	100.01
V	n.a.	398	213
Cr	38	49	4
Ni	n.a.	30	5
Rb	17	26	34
Sr	375	305	347
Y	n.a.	21	29
Zr	47	31	90
Nb	n.a.	n.a.	1
Ba	337	714	237
La	7	n.a.	42
Ce	16	n.a.	33
Th	1	n.a.	n.a.
Ta	n.a.	n.a.	n.a.
Hf	2	n.a.	n.a.

4.4.2   Regional Geology

The oldest rocks on the islands of Vanuatu are Oligocene submarine lavas (Coleman 1970). The arc was apparently rejuvenated in the Pliocene by a major period of submarine basaltic volcanism (Gorton 1977). In the Pleistocene and Recent, a third period of volcanism produced subaerial potassic olivine-basalts (Mitchell and Warden 1971; Gorton 1977).

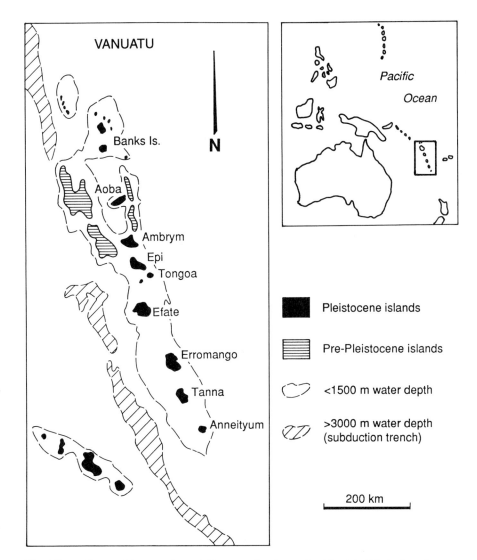

Fig. 4.8. Geographic overview of Vanuatu, Southwest Pacific. Modified after Gorton (1977).

4.4.3 Mineralogy and Petrography of the Potassic Igneous Rocks

The submarine volcanic rocks consist entirely of pillow lavas. They are characterized by porphyritic textures with phenocrysts of hypersthene, augite, plagioclase, and rare olivine in a glassy groundmass. Phenocrysts of biotite and amphiboles are normally absent (Gorton 1977).

The subaerial potassic olivine-basalts have porphyritic textures, with phenocrysts

Table 4.6.   Representative whole-rock major- and trace-element geochemistry of potassic igneous rocks from Vanuatu, Southwest Pacific. Major elements are in wt %, and trace elements are in ppm. Fe_2O_3 (tot) = total iron calculated as ferric oxide. From Gorton (1977).

Province/deposit:	Tongoa Island	Epi Island
Location:	Vanuatu	Vanuatu
Rock type:	Basalt	Andesite
Tectonic setting:	Late oceanic arc	Late oceanic arc
Reference:	Gorton (1977)	Gorton (1977)
SiO_2	49.42	60.74
TiO_2	0.60	0.86
Al_2O_3	13.14	15.24
Fe_2O_3 (tot)	10.99	7.64
MnO	0.20	0.13
MgO	9.37	2.92
CaO	12.77	4.94
Na_2O	1.76	3.59
K_2O	1.45	3.48
P_2O_5	0.30	0.46
LOI	0.64	1.28
Total	100.05	99.08
V	250	150
Cr	260	31
Ni	46	9
Rb	35	81
Sr	539	441
Y	17	34
Zr	57	176
Nb	1	4
Ba	160	1350
La	9	29
Ce	21	69
Th	2	5
Ta	n.a.	n.a.
Hf	n.a.	n.a.

of olivine, augite, and plagioclase in a feldspathic groundmass containing minor normative nepheline (Gorton 1977).

4.4.4 Geochemistry of the Potassic Igneous Rocks

The potassic olivine-basalts from Vanuatu are characterized by relatively high Al_2O_3 (up to 16 wt %), high CaO (up to 12.7 wt %), and high Na_2O (up to 3.5 wt %) contents (Gorton 1977): see Table 4.6. They have moderate mg# (< 70) and moderate concentrations of mantle-compatible elements (e.g. < 260 ppm Cr, < 250 ppm V) due to fractionation.

The very low concentrations of LILE (e.g. < 80 ppm Rb, < 540 ppm Sr), and low LREE (e.g. < 30 ppm La, < 70 ppm Ce) and HFSE (e.g. < 0.8 wt % TiO_2, < 170 ppm Zr) contents (Table 4.6) of the potassic rocks are typical for those derived in a late oceanic-arc setting (Müller et al. 1992b).

4.5 African Rift Valley (Rwanda, Uganda, Zaire): Example from a Within-Plate Setting

4.5.1 Introduction

The Virunga volcanic province (Fig. 4.9) occupies an area of more than 3000 km^2 at the junction of Rwanda, Zaire, and Uganda (De Mulder et al. 1986; Rogers et al. 1992). The highly potassic lavas of the province, in the western branch of the East African rift system, have long attracted interest (Thompson 1985). The Virunga igneous complex is widely regarded as one of the classic within-plate potassic provinces (Rogers et al. 1992, 1998). However, the question remains as to whether the rocks were derived from lithospheric mantle sources (De Mulder et al. 1986), or by

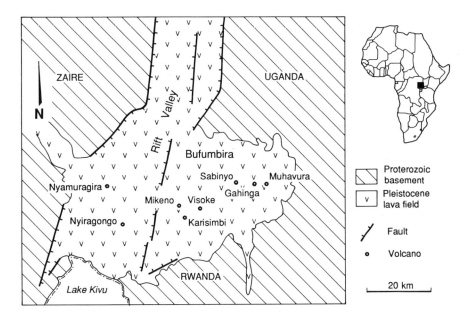

Fig. 4.9.   Geological overview of the Virunga volcanic province, African Rift Valley, Uganda. Modified after De Mulder et al. (1986).

the input of a deep asthenospheric mantle plume (Thompson 1985). The extreme compositions of some melilite-bearing lavas from Bufumbira seem to necessitate a high CO_2/H_2O ratio in their mantle sources, if this was lherzolite (Thompson 1985). These high CO_2 contents are indicative of asthenospheric mantle plumes (Thompson 1985; Nelson et al. 1986).

4.5.2 Regional Geology

Early Pliocene to late Pleistocene volcanic activity in the Virunga province produced eight major volcanoes (Fig. 4.9), two of which are still active (Rogers et al. 1992), and numerous minor vents (Cundari and Le Maitre 1970). The major volcanoes are situated in a large lava field. The volcanic activity is structurally related to graben-faulting during the development of a branch — known as the Bufumbira embayment — in the rift (De Mulder et al. 1986; Rogers et al. 1992). The Virunga volcanic rocks are underlain by metasedimentary rocks of the Karagwe-Ankolean System which was deposited about 2.1 Ga ago and deformed during the Kibaran Orogeny (1300–800 Ma; Vollmer and Norry 1983).

Karisimbi is the largest of the volcanoes and has been active during the last 0.1 Ma (Rogers et al. 1992). It was apparently formed during three phases, which are summarized by De Mulder et al. (1986) and Rogers et al. (1992). The initial phase formed a shield volcano comprising potassic basanites; this was followed by an eastward migration of volcanic activity and the development of a caldera complex. The second phase was characterized by potassic mugearite lavas. The final volcanic activity is manifested by highly potassic latite and trachyte lava flows.

4.5.3 Mineralogy and Petrography of the Potassic Igneous Rocks

The majority of the Virunga volcanic rocks are silica-deficient, leucite-bearing potassic rocks (Vollmer and Norry 1983). However, the Nyiragongo volcano also consists of nephelinites and melilitites (Demant et al. 1995). The potassic basanites, which have the most primitive compositions, are characterized by pronounced porphyritic or glomeroporphyritic textures with abundant phenocrysts of olivine, diopside, and chromite in a fine-grained or glassy groundmass comprising plagioclase, potassic feldspar or leucite, olivine, titaniferous salite, and titanomagnetite (Ferguson and Cundari 1975; Rogers et al. 1992). The mugearites consist mainly of potassic feldspar and diopside phenocrysts in a fine-grained groundmass of plagioclase, potassic feldspar, olivine, and diopside (De Mulder et al. 1986). Grading into the late-stage latites and trachytes, potassic feldspar joins plagioclase, biotite, and titanaugite as phenocrysts and commonly rims the plagioclase. Leucite cannot be optically identified in the groundmass (Ferguson and Cundari 1975).

Table 4.7.   Representative whole-rock major- and trace-element geochemistry of potassic igneous rocks from the Virunga volcanic field, Uganda. Major elements are in wt %, and trace elements are in ppm. Fe_2O_3 (tot) = total iron calculated as ferric oxide. From de Mulder et al. (1986).

Province/deposit:	Karisimbi, Virunga	Karisimbi, Virunga	Karisimbi, Virunga
Location:	Uganda	Uganda	Uganda
Rock type:	Basanite	Mugearite	Trachyte
Tectonic setting:	Within-plate	Within-plate	Within-plate
Reference:	De Mulder et al. (1986)	De Mulder et al. (1986)	De Mulder et al. (1986)
SiO_2	46.18	48.52	59.57
TiO_2	2.98	2.70	1.08
Al_2O_3	13.19	16.07	18.27
Fe_2O_3 (tot)	11.13	11.08	5.15
MnO	0.19	0.21	0.12
MgO	7.05	4.23	0.97
CaO	10.83	6.76	2.30
Na_2O	2.91	3.63	4.39
K_2O	3.44	4.48	6.87
P_2O_5	0.81	0.75	0.30
LOI	0.70	1.10	1.01
Total	99.41	99.53	100.03
V	n.a.	n.a.	n.a.
Cr	228	77	2
Ni	n.a.	n.a.	n.a.
Rb	115	151	242
Sr	1286	1280	698
Y	n.a.	n.a.	n.a.
Zr	311	341	414
Nb	60	n.a.	n.a.
Ba	1321	1445	1200
La	120	126	137
Ce	231	240	276
Th	20	26	42
Ta	9	10	10
Hf	8	8	10

4.5.4 Geochemistry of the Potassic Igneous Rocks

The whole-rock geochemistry of the Virunga potassic igneous rocks has been discussed by Mitchell and Bell (1976), De Mulder et al. (1986), and Rogers et al. (1992, 1998), and representative analyses are shown in Table 4.7.

Rogers et al. (1992) divided the rocks into two groups mainly based on their MgO contents. The older basanites which formed the shield volcano have primitive compositions with high MgO contents (> 6.3 wt %) and high mantle-compatible element concentrations (e.g. up to 1158 ppm Cr, up to 59 ppm Co; De Mulder et al.

1986). This can be compared to the mugearites with lower MgO contents (< 4.2 wt %) and lower mantle-compatible element concentrations (e.g. < 77 ppm Cr, < 33 ppm Co; De Mulder et al. 1986). The late-stage trachytes are characterized by more evolved compositions with very low MgO (< 1 wt %), Cr (< 2 ppm) and Co (< 5 ppm) contents. Crustal assimilation during uprise can be excluded based on the relatively high Ce/Pb ratios (> 10) of the Virunga rocks which are similar to those of MORB and OIB (Hofmann et al. 1986; Rogers et al. 1992). Many MORBs and OIBs have low Pb contents and high Ce/Pb ratios (~ 20) when compared to those melts derived from the continental crust (Ce/Pb < 6; Hofmann et al. 1986). The high HFSE concentrations (e.g. up to 4 wt % TiO_2, up to 91 ppm Nb, up to 18 ppm Hf; De Mulder et al. 1986; Rogers et al. 1992) of the Virunga rocks are typical for potassic igneous rocks derived in a within-plate setting (Müller et al. 1992b).

Stable-isotope studies by Rogers et al. (1992) indicate that the Virunga lavas were derived from the lithospheric mantle, and the evidence for a contribution from deep asthenospheric sources, as proposed by Thompson (1985), is very limited.

5 Primary Enrichment of Precious Metals in Potassic Igneous Rocks

5.1. Introduction

Controversy continues to surround the relative contributions of magmatic versus metamorphic and crustal versus mantle components to the fluids which are responsible for the origin of mesothermal gold deposits (e.g. Rock et al. 1989). The recognition that the products of deep-seated alkaline magmatism, such as lamprophyres, are spatially associated with many mesothermal gold deposits (Rock et al. 1989; Rock 1991) has resulted in the detailed study of primary precious-metal contents in these rocks (e.g. Wyman and Kerrich 1989a; Taylor et al. 1994). This chapter reviews some recent studies on primary precious-metal contents in shoshonitic and alkaline lamprophyres, and other potassic igneous rocks.

5.2 Theoretical Discussion

The occurrence of all primary PGE deposits in mafic–ultramafic intrusions points to a congruent enrichment of precious metals in parts of the mantle (Rock et al. 1988a). Cabri (1981) considers most of the terrestrial precious-metal budget to have been partitioned into the deep mantle and core during the early differentiation of the planet. The ideal magma-type to transport these elements into the crust is, therefore, one with an ultrabasic character and an exceptionally deep origin (i.e. lamprophyres; cf. Rock et al. 1988a). Lamprophyres also have magmatic compositions (high CO_2 and halogen contents) potentially suitable for transporting Au from the mantle into the crust (Rock et al. 1988a; Rock 1991).

Recent reviews have shown that some mantle-derived lamprophyric melts have very high primary PGE contents (e.g. Crocket 1979). A compilation of precious-metal contents of potassic igneous rocks, such as lamprophyres and related rocks (Table 5.1), has been given by Rock et al. (1988a, 1989). For example, lamproites from the Ellendale Field, Western Australia, contain up to 56 ppb Pd (Lewis 1987). The Wessleton and Frank Smith kimberlites, South Africa, contain up to 19 ppb Pd

Table 5.1. Compilation of precious metal abundances in potassic igneous rocks. NAA = neutron activation analysis, ICP = inductively coupled plasma mass spectrometry, n.a. = not analyzed. After Rock et al. (1988a, 1989).

Individual intrusion	Reference	Method	Au (ppb)	Pt (ppb)	Pd (ppb)
Kimberlites					
South Africa					
Wessleton	Paul et al. (1979)	NAA	n.a.	n.a.	18
Frank Smith	Paul et al. (1979)	NAA	n.a.	n.a.	18
Lamproites					
Kimberley, Australia					
Ellendale 9	Lewis (1987)	ICP	n.a.	3	1
Ellendale 11	Lewis (1987)	ICP	n.a.	4	56
Lamprophyres					
Borneo					
Linhaisai minette	Bergman et al. (1988)	NAA	15	n.a.	n.a.
Canada					
Malpeque	Greenough et al. (1988)	NAA	45	n.a.	n.a.
Papua New Guinea					
Fu lamprophyre	Finlayson et al. (1988)	NAA	29	n.a.	n.a.

(Paul et al. 1979), and discrete grains of platinum-group metals have been detected in these kimberlites (Mitchell 1986). The precious-metal concentrations in lamproites and kimberlites, which normally occur in within-plate settings and represent the deepest forms of magmatism, suggest the presence of precious-metal-enriched source regions within the upper mantle (Rock et al. 1988a). Their precious-metal enrichments are interpreted to be primary magmatic features (see below), since lamproites and kimberlites are rapidly irrupted through the crust and commonly show little evidence of significant fractionation during ascent, as reflected in their high mg# and the occurrence of mantle xenoliths (Rock et al. 1988a; Rock 1991).

The association in space and time between shoshonitic lamprophyres and mesothermal gold deposits has been documented in many Archaean greenstone-belt terrains (Hallberg 1985; Taylor et al. 1994). Platinum-group elements such as Ir, Os, and Ru are mantle compatible, and they remain in the olivine (forsterite)-bearing residue of the mantle during partial melting (Brügmann et al. 1987; Gueddari et al. 1996). Originally sulphur-rich magmas such as MORB are generally PGE-poor, because the sulphide-hosted PGE are removed during fractionation. In contrast, the elements Cu, Au, Pt, and Pd behave incompatibly during partial melting of a strongly depleted mantle source (Brügmann et al. 1987; Stanton 1994) forming sulphur-undersaturated melts (Hamlyn et al. 1985). Copper, Au, Pt, and Pd behave as mantle-incompatible elements (Gueddari et al. 1996) and are partitioned into the first silicate-melt increments, rather than into a separate sulphide-liquid fraction (Taylor et al. 1994). Fractionation of these precious-metal-enriched, but sulphur-undersaturated, parent magmas can lead to further Au and PGE enrichment provided the melt does not become sulphur saturated (Hamlyn et al. 1985; Brügmann et al. 1987; Taylor et al. 1994). Gold probably behaves incompatibly during olivine fractionation, because the Au^{2+} oxidation state is not known from natural systems (Togashi and Terashima 1997). Native Au, Au^+, or Au^{3+} are the most stable species, but their charges and their large ionic radii preclude their partitioning into olivine (Brügmann et al. 1987).

Under conditions of relatively high fO_2, sulphide species become unstable with respect to sulphates in the magma and the segregation of an immiscible sulphide-melt phase is impossible (Richards 1995). Under such conditions, the chalcophile elements behave as incompatible components in the melt, becoming steadily enriched during fractionation (D. Wyborn, pers. comm. 1995). Additionally, the high fO_2 may delay sulphur saturation until after the onset of volatile saturation, thus generating chalcophile-enriched hydrothermal fluids (Richards 1995). Hence, the observed Au and PGE enrichments of many potassic lamprophyres (see Sect. 5.3) could be primary features. Trace-element data for potassic igneous rocks such as shoshonites suggest that they are also enriched in Cu (Kesler 1997). The fact that not all potassic igneous rock suites are mineralized, and that not all shoshonitic lamprophyres are Au- or PGE-enriched (see Chap. 7), probably indicates the heterogeneity of mantle metasomatism processes (Taylor et al. 1994). It is not yet clear whether the oxidized nature of late oceanic-arc basalts is a source characteristic or whether it is caused by secondary processes such as degassing (Ballhaus 1993). For instance, the rapid loss of hydrogen by diffusion, with an attendant increase in the H_2O/H_2 ratio, increases fO_2 (Haggerty 1990), and has been modelled as the mechanism to account for oxidized crusts in ponded lava lakes (Sato and Wright 1966). Late oceanic-arc basalts, which are commonly more enriched in volatiles than other basalts (Muenow et al. 1990), are more susceptible to those degassing processes.

Generally, there are two mechanisms for the generation of sulphur-undersaturated precious-metal-rich magmas (Sun et al. 1991):

- High temperature (> 1400°C), large degree ($\geq 25\%$) of mantle melting related to asthenospheric mantle plume activity.
- Lower temperature, small degree of partial melting of mantle wedge material ($\leq$ 4 ppb Pt, Pd; $\leq$ 1 ppb Au; < 250 ppm S) in subduction arcs.

The second process might produce potassic lamprophyres and shoshonites with primary precious-metal enrichments (S.-S. Sun, pers. comm. 1993). However, too small a degree of partial melting (< 10%) will leave sulphides in the mantle residue, and hence retain the precious metals during such partial melting (Sun et al. 1991). As noted above, the conditions in subduction zones are considered to be too oxidizing for the generation of sulphur-saturated magmas (Taylor et al. 1994).

Silicate minerals have low mineral/melt distribution coefficients for Pd, Pt, and Au (Keays 1982), when compared to the very high partition coefficients for PGE into sulphides (Campbell et al. 1983). Thus, the PGE tend to form sulphides if sufficient sulphur is present in the melt. These relatively dense sulphides are strongly affected by gravitational fractionation, which means that the resulting melt will be gradually depleted in precious metals en route to surface. As a consequence, efficient precious-metal concentration in a magma chamber (Sun et al. 1991) requires fertile mafic magmas with high PGE background levels (e.g. > 15 ppb Pd), but low sulphur contents (< 1000 ppm S).

Obvious exceptions to the rule regarding correlation between sulphur-undersaturation and precious-metal abundances in mafic magmas are the sulphur-saturated lamproites and kimberlites which may contain significant precious-metal concentrations (Mitchell and Keays 1981). However, this can be explained by the volatile-driven rapid uprise of these potassic magmas, which might sample some precious-metal-enriched mantle sulphide droplets, in addition to mantle xenoliths, en route to surface (Sun et al. 1991).

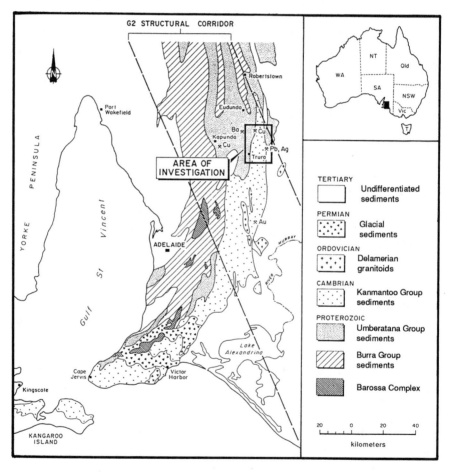

Fig. 5.1.   Geological overview of the southern part of the Adelaide Geosyncline, South Australia, showing the study area, which includes the Karinya Syncline. Modified after Müller et al. (1993a).

Theoretically, on the basis of the discussion above, primary Au and PGE enrichment of lamprophyric magmas should be characterized by elevated concentrations of all incompatible metals such as Cu, Au, Pt, and Pd. If Au peaks are decoupled from Cu, Pt, and Pd peaks in distribution plots of metal contents in the lamprophyres normalized to primitive mantle, it is likely that the anomalous Au contents are secondary features (Wyman and Kerrich 1989a).

5.3   Case Study: Potassic Alkaline Lamprophyres with Elevated Gold Concentrations from the Karinya Syncline, South Australia

5.3.1   Introduction

During recent extensive base-metal exploration in the Karinya Syncline, which forms part of the Adelaide Geosyncline in South Australia, a lamprophyre province has been mapped in the area between Truro and Frankton about 80 km northeast of Adelaide (Fig. 5.1; Morris 1990; Müller et al. 1993a). The investigated dykes from the Karinya Syncline have been classified as alkaline lamprophyres (Müller et al. 1993a), but they also show transitional features to lamproites. Potassic lamprophyres with lamproitic affinity are known from several localities worldwide (e.g. Wagner and Velde 1986; Venturelli et al. 1991a), but their petrogenesis is still poorly understood (Foley et al. 1987).

The potassic lamprophyres from the mineralized Karinya Syncline have high Au contents. As they have also been analyzed for other base- and precious-metals (Cu, Ni, Pt, Pd) and for elements commonly associated with hydrothermal gold deposits (As, Sb, W), it is informative to use them as a test of primary Au enrichment of these rocks as proposed, for example, by Rock and Groves (1988a, 1988b). However, before doing this, the mineralogy, petrology and geochemistry of the suite must be documented to determine whether the rocks are altered or not. It is also informative to use the geochemical data as a test of the tectonic discrimination diagrams presented in Chapter 3. A more complete description is provided by Müller et al. (1993a).

5.3.2   Regional Geology and Tectonic Setting

The Adelaide Geosyncline is a 700 km-long fold belt comprising late Proterozoic to middle Cambrian sedimentary rocks which were folded, metamorphosed, and uplifted during the Delamerian Orogeny in the late Cambrian (Thomson 1969, 1970). No reliable indicators for an ancient subduction event (i.e. blueschists, ophiolites, mélanges) have been recorded in the Adelaide Geosyncline, and Preiss (1987) has suggested a within-plate origin for the igneous activity, perhaps related to deep-seated crustal fractures.

The Karinya Syncline, forming the northern part of the Kanmantoo Trough (southern part of the Adelaide Geosyncline), consists of Cambrian metasedimentary rocks which were intruded by the lamprophyre swarm during the Ordovician (Fig. 5.2).

5.3.3 Mineralization in the Vicinity of the Lamprophyres

In the Adelaide Geosyncline, mineral deposits of various kinds and different ages are situated either in, or at the edges of, major north-northwest-trending lineaments or at their intersections (O'Driscoll 1983). O'Driscoll (1983) has shown that the

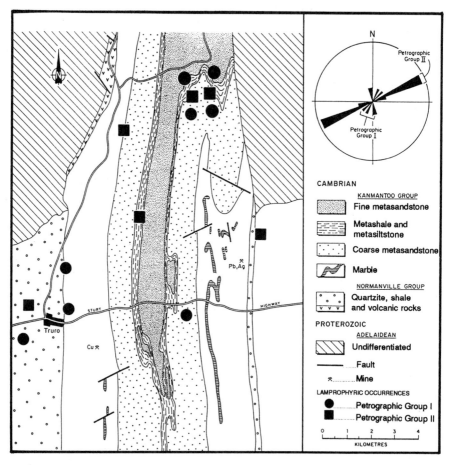

Fig. 5.2.   Geological overview of the investigated area within the Karinya Syncline, South Australia. The map shows lamprophyre localities and two former base-metal mines in the area; circles represent samples from petrographic group 1 and squares represent those from petrographic group 2. The rose diagram shows the different strike directions of the two distinctive petrographic groups. Modified after Müller et al. (1993a).

Table 5.2.   Whole-rock K-Ar dating on lamprophyres from the Karinya Syncline, South Australia. Analyses performed by Analabs Laboratories, Adelaide, South Australia. The error on ages is quoted as ± 2σ. Sample numbers refer to specimens held in the Museum of the Department of Geology and Geophysics, The University of Western Australia. From Müller et al. (1993a).

Sample no.	Petrographic group	K (wt %)	K-Ar age (Ma)
119068	I	7.33	458 ± 2
119072	II	7.84	478 ± 3
119078	II	8.29	481 ± 3

largest lineaments coincide with small base-metal deposits at Kanmantoo and Kapunda, as well as with the giant copper-gold-uranium deposit at Olympic Dam (Creaser 1996). These deposits are, however, older than the lamprophyres discussed here. The area also hosts several smaller gold mines, where shallow alluvial gold-fields and some reefs were developed (e.g. Moppa and Hamilton areas south of Truro), as well as barite deposits (e.g. northwest of Dutton, west of Truro; Horn et al. 1989).

The occurrence of mineral deposits, in combination with the presence of crustal-scale lineaments which represent zones of deep-seated crustal weakness, reflects the potentially high prospectivity of the area for further gold and base-metal discoveries (cf. Rock et al. 1988a), although to date there are no clear indications of mineralization directly related to the lamprophyres.

5.3.4 Nature of the Lamprophyres

Most of the dykes strike northeasterly, as shown in Figure 5.2. The thickness of the dykes varies from 0.1 to 1.5 m, and, in general, they have intruded joint planes of the country rock. The lamprophyre dykes show typical porphyritic textures, with phlogopite phenocrysts in a fine-grained groundmass. Most lamprophyre dykes show flow textures, with a parallel alignment of phlogopite phenocrysts, and several have chilled margins at contacts with their host rocks. In the Truro area there is also a lamproitic diatreme. All samples are affected to varying extents by secondary altera-tion. The carbonate host rocks at Robertstown are strongly altered to talc, asbestos, and tourmaline.

The investigated lamprophyres were emplaced at a shallow depth during the Ordovician after the Delamerian Orogeny. Two different K-Ar ages of 458 ± 2 Ma and 480 ± 3 Ma (Table 5.2) are consistent with two distinctive petrographic groups of lamprophyres (see below) as discussed by Müller et al. (1993a).

The large (1–6 mm), euhedral phlogopite phenocrysts are sited in a groundmass of mainly felsic components (e.g. orthoclase, leucite, plagioclase, quartz). The phlogopites commonly show battlement structures (Fig. 5.3c) and parallel orientations (i.e. flow textures; Fig. 5.3a–b). An older phenocryst generation is represented by large (up to 6 mm), zoned, Na-rich phlogopite crystals, and a younger generation by smaller (< 1 mm) phlogopite crystals. Several rocks also have apatite and titanite

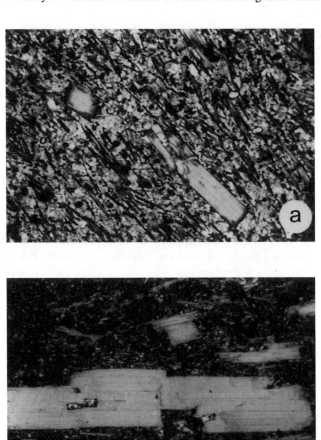

Fig. 5.3.   Photomicrographs (crossed nicols) of typical lamprophyre samples from the Karinya Syncline, South Australia. (a) Two generations of mica phenocrysts with varying size (119068) [FOV 4 mm]. (b) Flow textures displayed by phlogopite phenocrysts (119072) [FOV 2 mm]. (c) Battlement structures in phlogopite phenocryst (119068) [FOV 2 mm]. Sample numbers refer to specimens held in the Museum of the Department of Geology and Geophysics, The University of Western Australia.

microphenocrysts.

Based on phenocryst mineralogy, two petrographic types of lamprophyres can be distinguished, one is *phlogopite phyric* and the other is *apatite-phlogopite phyric*. Primary ferromagnesian phenocrysts other than phlogopite are generally absent. However, one sample also contains alkali-amphibole (i.e. riebeckite), and others show secondary amphibole (i.e. cummingtonite) resulting from intensive alteration. Phenocryst mineralogy is dominated by large zoned phlogopites with very high mg# of 89–90, variable TiO_2 contents (1.71–4.06 wt %), and Al_2O_3 contents between 12 and 13 wt % (cf. Rock 1991). The phlogopite phenocrysts are characterized by relatively high F concentrations (up to 3.79 wt %) compared to average values of typical lamproites (Table 5.3). For example, Miocene olivine-lamproites from the West Kimberley of Australia have F concentrations of only about 1.71 wt % (Jaques et al. 1986; Mitchell and Bergman 1991) and Cretaceous lamproites from Prairie Creek, Arkansas, have F values below 0.94 wt % (Scott-Smith and Skinner 1982; Mitchell and Bergman 1991). Mica compositions, plotted on a Al-Mg-Fe triangular diagram (Mitchell and Bergman 1991; Sheppard and Taylor 1992), show some overlap between values typical for lamproites and lamprophyres (Fig. 5.4a).

Groundmass minerals are mainly orthoclase, leucite, plagioclase, and quartz. Potassic feldspars, plotted on a Fe_2O_3 versus orthoclase biaxial diagram (e.g. Sheppard and Taylor 1992), show transitional features between lamproites and alkaline lamprophyres (Fig. 5.4b). Lamproites (sensu stricto), which commonly show leucite, do not contain plagioclase (Velde 1975; Bergman 1987). They are also characterized by the presence of Cr-spinels, which have not been detected in the described lamprophyres.

Several dykes have an extremely fine-grained, cryptocrystalline, and partly glassy groundmass. Two lamprophyres show syenitic ocelli, mainly consisting of orthoclase, which are irregular shaped globular structures and gradational with their host-rocks (cf. Rock et al. 1988b; Perring et al. 1989b; Rock 1991).

Importantly for the study of precious-metal concentrations, some samples are affected by secondary alteration. Alteration consists mainly of saussuritization of groundmass feldspars producing secondary epidote, or secondary carbonate replacement. Three samples are strongly altered, forming magnesio-cummingtonites with a fibrous character and anomalous birefringence; their felsic groundmass minerals are completely altered to secondary talc and epidote.

Table 5.3.   Microprobe (wavelength dispersive spectrum - WDS) analyses of mica phenocrysts from lamprophyres from the Karinya Syncline, South Australia. FeO (tot) = total iron
calculated as ferrous oxide, Cumm. = cummingtonite, Ox. form. = oxygen formula. Sample
numbers refer to specimens held in the Museum of the Department of Geology and Geophysics, The University of Western Australia. From Müller et al. (1993a).

Sample no.: Petrographic group:	119068 I	119072 II	119073 II	119074 II
wt %				
SiO_2	39.58	40.24	37.51	40.07
TiO_2	2.21	2.18	4.06	1.71
Al_2O_3	13.35	12.46	12.94	11.89
Cr_2O_3	0.10	0.13	0.02	0.42
FeO (tot)	5.87	6.15	12.42	6.35
MnO	0.06	0.05	0.10	0.05
MgO	22.79	22.83	17.34	22.28
CaO	0.03	0.03	0.04	0.03
Na_2O	0.32	0.12	0.21	0.08
K_2O	10.09	10.22	9.74	10.31
SrO	0.11	0.12	0.09	0.10
BaO	0.28	0.46	0.65	0.31
Cl	0.03	0.02	0.02	0.02
F	1.32	1.55	0.83	3.24
Total	95.56	95.89	95.60	95.48
Name	Phlogopite	Phlogopite	Phlogopite	Phlogopite
Mg#	90	89	76	89
Ox. form.	22	22	22	22
Atoms				
Si	5.728	5.823	5.612	5.895
Ti	0.241	0.237	0.456	0.189
Al	2.277	2.124	2.282	2.061
Cr	0.011	0.015	0.001	0.049
Fe	0.710	0.744	1.553	0.781
Mn	0.007	0.005	0.012	0.003
Mg	4.912	4.921	3.864	4.883
Ca	0.002	0.005	0.007	0.002
Na	0.090	0.034	0.059	0.023
K	1.863	1.885	1.858	1.934
Sr	0.007	0.010	0.005	0.008
Ba	0.016	0.026	0.038	0.018
Cl	-	-	-	-
F	-	-	-	-
Total	15.864	15.829	15.749	15.847

119076 I	119079 II	119080 I	119070 I	119081 I
39.86	39.54	40.36	40.04	52.90
1.68	2.15	1.79	1.73	5.32
11.41	12.14	12.30	14.25	0.42
0.17	0.33	0.18	0.26	0.02
6.01	6.22	6.02	11.20	17.16
0.05	0.06	0.05	0.27	0.05
23.33	22.23	22.52	28.42	12.77
0.02	0.03	0.03	0.56	0.80
0.29	0.15	0.26	0.59	6.71
10.04	10.46	9.51	0.35	0.42
0.09	0.11	0.09	0.12	0.18
0.25	0.36	0.33	0.13	0.11
0.02	0.02	0.02	0.09	0.02
3.79	2.76	3.02	0.40	0.27
95.40	95.38	95.19	98.15	97.02
Phlogopite	Phlogopite	Phlogopite	Cumm.	Riebeckite
90	89	90	85	63
22	22	22	23	23
5.943	5.817	5.900	5.720	7.811
0.184	0.237	0.197	0.186	0.591
1.956	2.104	2.118	2.400	0.074
0.019	0.038	0.021	-	0.001
0.730	0.765	0.735	1.338	2.118
0.005	0.007	0.007	-	0.002
5.053	4.870	4.904	6.052	2.809
0.002	0.004	0.001	0.085	0.127
0.081	0.042	0.072	0.163	1.919
1.862	1.962	1.772	0.063	0.079
0.008	0.009	0.007	0.011	-
0.014	0.021	0.019	-	0.006
-	-	-	-	-
-	-	-	-	-
15.857	15.877	15.753	16.018	15.537

5.3.5 Petrology and Geochemistry of the Lamprophyres

The major- and trace-element chemistry of representative samples is given in Tables 5.4 and 5.5. The dykes show SiO_2 contents between 39.6 and 63.1 wt % (Fig. 5.5) and mg# varies from 35 to 73.

Most samples have high K_2O (> 3 wt %) and MgO (> 3 wt %) contents, and K_2O/Na_2O ratios > 2, and can be classified as ultrapotassic (Foley et al. 1987). However, the most-altered samples have relatively low K_2O contents (< 1.23 wt %) due to mobilisation of K during alteration. The lamprophyres are characterized by very high F concentrations (1050–5800 ppm).

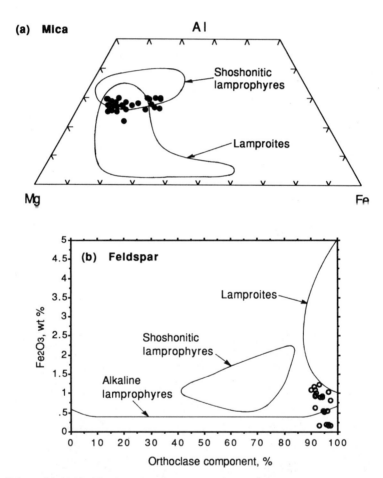

Fig. 5.4.   (a) Al-Mg-Fe triangular plot showing 47 representative mica analyses from the Karinya Syncline, South Australia. (b) Fe_2O_3 versus orthoclase biaxial plot showing 21 representative potassic feldspar analyses for lamprophyres from the Karinya Syncline, South Australia. The orthoclase component of the investigated feldspar phases is given in %. Data from Müller et al. (1993a). Modified from Sheppard and Taylor (1992).

Table 5.4.   Selected major-element analyses (in wt %) of lamprophyres from the Karinya Syncline, South Australia. Sample numbers refer to specimens held in the Museum of the Department of Geology and Geophysics, The University of Western Australia. From Müller et al. (1993a).

	Fresh samples		Altered samples	
Sample no.:	119067	119072	119070	119078
Petrographic group:	I	II	I	II
SiO_2	56.30	56.00	40.01	48.00
TiO_2	2.02	1.86	1.82	1.75
Al_2O_3	12.29	12.10	12.84	9.50
Fe_2O_3 (tot)	9.59	8.50	13.03	9.81
MnO	0.02	0.06	0.02	0.25
MgO	4.83	6.70	17.96	6.20
CaO	0.40	2.08	0.42	6.68
Na_2O	0.30	1.42	1.51	0.39
K_2O	9.10	7.55	1.08	7.69
P_2O_5	0.48	1.60	0.31	1.80
LOI	3.98	2.13	10.71	6.57
Total	99.31	100.00	99.70	98.64
Mg#	50	61	73	56
(Na+K)/Al	0.84	0.87	0.28	0.94

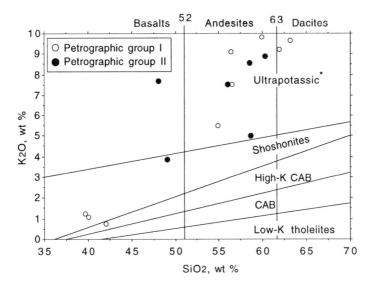

Fig. 5.5. K_2O versus SiO_2 plot (after Peccerillo and Taylor 1976a) showing the potassic or ultrapotassic chemistry of the lamprophyres from the Karinya Syncline, South Australia. * = ultrapotassic, as defined by Foley et al. (1987); CAB = calc-alkaline basalts. From Müller et al. (1993a).

Table 5.5.   Selected trace-element analyses of lamprophyres from the Karinya Syncline, South Australia. Trace elements are in ppm, and precious metal are in ppb. n.a. = not analyzed. Sample numbers refer to specimens held in the Museum of the Department of Geology and Geophysics, The University of Western Australia. From Müller et al. (1993a).

	Fresh samples		Altered samples	
Sample no.:	119067	119072	119070	119078
Petrographic group:	I	II	I	II
F	4300	5500	4500	n.a.
Li	62	123	62	43
Sc	20	17	33	16
V	587	544	327	502
Cr	70	105	145	100
Co	10	40	55	25
Ni	120	265	345	185
Cu	160	35	65	300
Zn	90	205	155	155
As	13	9	2	2
Rb	575	496	10	411
Sr	614	451	91	740
Y	89	64	36	27
Zr	777	810	165	818
Nb	26	28	11	20
Sb	1.4	2.0	0.2	3.2
Ba	2878	2494	165	2538
La	57.5	51.6	6.9	83.5
Ce	120	100	16	157
Nd	50	40	9	56
Sm	8.9	7.3	2.5	8.8
Yb	3.5	3.4	2.8	1.1
Hf	18	21	3.5	19
W	3.0	1.2	1.7	2.5
Pb	5	59	2	114
Pd	13	9	5	49
Pt	< 5	< 5	< 5	19
Au	12	4	< 3	23

Dykes of petrographic type 2 show P_2O_5 values > 1.4 wt %, whereas those of petrographic type 1 are characterized by P_2O_5 values < 1.4 wt %. Petrographic type 2 dykes show distinctive Nb concentrations between 20 and 30 ppm, whereas those of type 1 have very variable Nb concentrations, between 10 and 70 ppm. However, distinction between the two petrographic groups using spidergram patterns is equivocal.

The relatively low mg# of the rocks suggest that the lamprophyres are derived from primitive melts via olivine ± clinopyroxene ± phlogopite ± apatite fractionation. This is consistent with their relatively low Ni concentrations. (Na+K)/Al ratios show that the samples tend towards peralkaline character (Table 5.4). The relatively

Table 5.6.   Correlation matrix for precious metals (Au, Pd, Pt), Cu, and gold pathfinder elements (As, Sb, W) of lamprophyres from the Karinya Syncline, South Australia.

	Cu	Au	Pt	Pd	As	Sb	W
Cu	1						
Au	0.772	1					
Pt	0.73	0.604	1				
Pd	0.786	0.9	0.849	1			
As	-0.14	-0.188	-0.363	-0.254	1		
Sb	0.41	0.136	0.05	0.115	0.316	1	
W	-0.234	-0.238	-0.088	-0.22	0.264	0.117	1

high TiO_2 contents of the investigated lamprophyres (up to 2.36 wt %) and their high Zr concentrations (384–1135 ppm) reflect their alkaline geochemistry (see Sect. 4.2; Müller et al. 1992a). Two samples have very low SiO_2 contents (< 41 wt %) and very high MgO contents (> 16 wt %), suggesting an affinity to lamproites. However, despite their high MgO contents, they have only relatively low Ni (< 345 ppm), Cr (< 145 ppm), and Ba (< 165 ppm) contents. This clearly distinguishes them from typical olivine-lamproites (Müller et al. 1993a), and suggests that the high MgO contents are not primary, but caused by secondary alteration processes.

The geochemistry of the investigated lamprophyres is atypical of lamproites (cf. Mitchell and Bergman 1991) because of their unusually low LREE abundances (e.g. < 147 ppm La, < 287 ppm Ce). The lamprophyres from the Karinya Syncline also have relatively high Al_2O_3 (~ 12 wt %) and low Ba contents (~ 2800 ppm in fresh samples) compared with average values for lamproites of 4–10 wt % and 1–3 wt %, respectively (Bergman 1987).

In order to determine the tectonic setting of the investigated potassic lamprophyres, they were plotted on the TiO_2 versus Al_2O_3, Y versus Zr, and Zr/Al_2O_3 versus TiO_2/Al_2O_3 biaxial discrimination plots of Müller et al. (1992b). The fresh potassic lamprophyres from the Karinya Syncline are characterized by very high LILE, LREE, and HFSE concentrations, and plot in the fields of within-plate types in Figure 5.6. This implies that the rocks were generated in a within-plate setting, consistent with previous tectonic interpretations of the area (Preiss 1987).

5.3.6 Precious Metal Abundance and Significance

Potassic igneous rocks are established as being closely related with certain types of gold and base-metal deposits (Mitchell and Garson 1981). Previous work shows that lamprophyres, too, can contain elevated concentrations of precious metals, as discussed above. As documented in Table 5.5, the lamprophyres from the Karinya Syncline contain up to 23 ppb Au, up to 19 ppb Pt, and up to 49 ppb Pd. These are well above the normal background levels of these elements for basic igneous rocks, which are commonly less than 2 ppb (Taylor et al. 1994). Figure 5.7 shows that Au enrichment of the South Australian lamprophyres is not decoupled from Cu and Pd

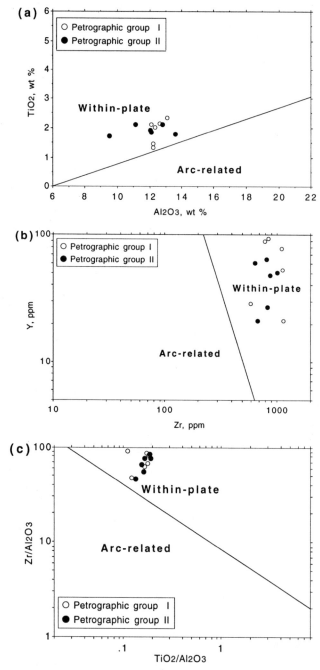

Fig. 5.6. Discrimination diagrams for potassic igneous rocks (see Chap. 3) indicating a within-plate tectonic setting for the investigated lamprophyres. (a) TiO_2 versus Al_2O_3 plot. (b) Y versus Zr plot. (c) Zr/Al_2O_3 versus TiO_2/Al_2O_3 plot. From Müller et al. (1993a)

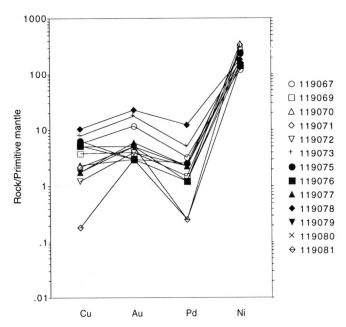

Fig. 5.7.   Abundances of chalcophile elements in lamprophyres from the Karinya Syncline, South Australia, relative to primitive mantle. Normalizing factors after Brügmann et al. (1987).

peaks in primitive mantle-normalized distribution plots (after Brügmann et al. 1987), suggesting that the anomalous Au contents *are* primary features (Wyman and Kerrich 1989a). The high correlation (> 0.6, < 0.9) of the elements Cu, Au, Pd, and Pt is also shown in the correlation matrix for these elements, further supporting a primary precious-metal enrichment of the lamprophyres (Table 5.6). A precious-metal enrichment of the dykes by crustal assimilation during uprise (see Sect. 4.2) is improbable, and enrichment by hydrothermal fluids after their emplacement (see Sect. 7.2) seems unlikely because there is a poor correlation (< 0.2) between Au and the pathfinder elements for hydrothermal mineralization (Table 5.6). Thus, the lamprophyres appear to be examples where there is a primary enrichment of precious metals (cf. Rock and Groves 1988a, 1988b; Rock et al. 1988a).

5.4   Comparison of Precious Metal Abundances for Lamprophyres from the Karinya Syncline and Kreuzeck Mountains

This section compares the precious-metal abundances of potassic lamprophyres from the Karinya Syncline, South Australia (Sect. 5.3) with those from the Kreuzeck

Table 5.7.   Correlation matrix for precious metals (Au, Pd, Pt), Cu, and gold pathfinder elements (As, Sb, W) of dykes from the Kreuzeck Mountains, Austria.

	Cu	Au	Pt	Pd	As	Sb	W
Cu	1						
Au	-0.043	1					
Pt	-	-	1				
Pd	0.229	0.142	-	1			
As	0.229	-0.17	-	-0.125	1		
Sb	-0.125	0.519	-	0.022	-0.035	1	
W	-0.028	0.473	-	-0.025	-0.289	-0.138	1

Mountains, Eastern Alps, Austria (Sect. 4.2). The Alpine lamprophyres are also characterized by elevated precious-metal concentrations (Table 4.3) but, importantly, these are not primary features, as discussed below.

All lamprophyre samples from the Kreuzeck Mountains, Eastern Alps, were analyzed for Au and PGE using Pb-fire-assay and graphite-furnace atomic-absorption spectroscopy (AAS) techniques at the Institute of Geology, Mining University, Leoben, Austria. Samples were prepared for analysis according to the method of Sighinolfi et al. (1984). The detection limit was 3 ppb for Au, 5 ppb for Pt, and 1 ppb for Pd, and the results were checked against a SARM-7 standard.

The Kreuzeck lamprophyres contain up to 27 ppb Au and up to 19 ppb Pd (cf. Müller et al. 1992a), with one calc-alkaline basaltic dyke containing 34 ppb Au. This can be compared with the 0.5–2 ppb Au content that is the typical background level for basic igneous rocks (Taylor et al. 1994). The Au-bearing samples from the Kreuzeck Mountains are mainly exposed in the central part of the area, where massive sulphides with significant Au and Ag concentrations were mined. The lamprophyres enriched in Au normally show Na_2O contents of more than 2.30 wt %, and commonly have high LOI.

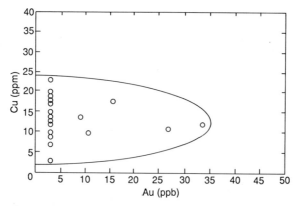

Fig. 5.8.   Plot of Cu versus Au for lamprophyres from the Kreuzeck Mountains, Austria.

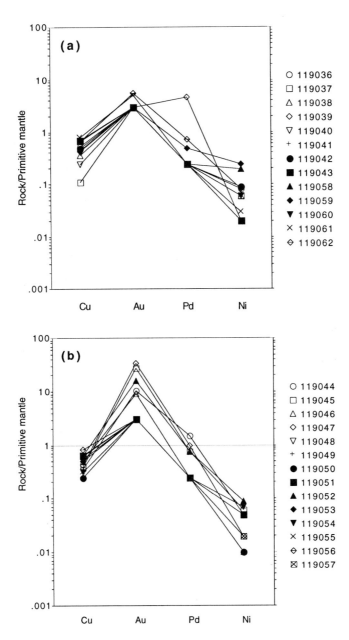

Fig. 5.9.   Abundances of chalcophile elements in dykes from the Kreuzeck Mountains, Austria, relative to the primitive mantle. Normalizing values after Brügmann et al. (1987). (a) Dykes from the northern and southern parts of the area. (b) Dykes from the central parts of the area where most former gold mines are located.

A Cu versus Au biaxial plot for these rocks (Fig. 5.8) shows no linear correlation between the two elements as would be expected if the magmas had a primary metal enrichment (Taylor et al. 1994). Figure 5.9 shows that elevated Au values of the dykes are decoupled from Cu, Pt, and Pd values in distribution plots of metal contents normalized to primitive mantle (after Brügmann et al. 1987), contrasting with data for the lamprophyres from the Karinya Syncline (Fig. 5.7). Again, this suggests that the anomalous Au contents are secondary features (cf. Wyman and Kerrich 1989a). Figure 5.9b shows that the Au concentrations of dykes from the central parts of the Kreuzeck Mountains are slightly higher than those of dykes from the northern and southern parts of the area (Fig. 5.9a). Table 5.7 shows a correlation matrix for precious metals (Au, Pt, Pd), Cu, and pathfinder elements (As, Sb, W) for hydrothermal gold deposits, illustrating the weak correlation between these elements. If the elevated precious-metal contents in some of the lamprophyres were direct enrichments related to hydrothermal gold mineralization, a better correlation between gold and its normal pathfinder elements would be expected. Therefore, the available data support neither a primary magmatic enrichment model nor a hydrothermal enrichment model. Müller et al. (1992a) suggest that, in this case, the high Au and PGE contents are related to assimilation of the Au-enriched massive-sulphide deposits in the area during uprise and emplacement of the lamprophyric magmas.

The above interpretation is consistent with recent studies of shoshonitic lamprophyres from the Hillgrove gold-antimony mining district, New South Wales, Australia (Ashley et al. 1994), where the lamprophyres are enriched in the elements of the mineralization that they cut.

6 Direct Associations Between Potassic Igneous Rocks and Gold-Copper Deposits

6.1 Direct Associations in Specific Tectonic Settings: Introduction

An overview of spatial associations between potassic igneous rocks and gold-copper deposits in the Southwest Pacific area is shown in Figure 6.1. Examples of direct associations between potassic igneous rocks and copper-gold deposits investigated in this study are, in order of increasing age:

- The Quaternary Ladolam gold deposit, Lihir Island, Papua New Guinea (Wallace et al. 1983; Plimer et al. 1988; Moyle et al. 1990; Dimock 1993; Hoogvliet 1993).
- The Tertiary Emperor gold deposit, Viti Levu, Fiji (Gill 1970; Colley and Greenbaum 1980; Anderson and Eaton 1990).
- The Pliocene Grasberg gold-copper deposit, Papua New Guinea (MacDonald and Arnold 1994).
- The Pliocene Misima gold deposit, Misima Island, Papua New Guinea (Lewis and Wilson 1980; Wilson and Barwick 1991; Appleby et al. 1995).
- The Miocene Porgera gold deposit, Papua New Guinea (Handley and Henry 1990; Richards 1990a, 1990, 1992; Richards et al. 1990, 1991).
- The Miocene Bajo de la Alumbrera copper-gold deposit, Catamarca Province, Argentina (Müller and Forrestal 1998).
- The Miocene Dinkidi copper-gold deposit, Didipio, Philippines (Haggman 1997a, 1997b; Garrett 1999).
- The Miocene El Indio gold deposit, Chile (Jannas et al. 1990).
- The Eocene Bingham copper deposit, Utah, USA (Lanier et al. 1978a, 1978b; Warnaars et al. 1978; Bowman et al. 1987).
- The Paleocene Twin Buttes copper deposit, Arizona, USA (Titley 1982; Barter and Kelly 1982).
- The Ordovician Goonumbla copper-gold deposit, New South Wales, Australia (Heithersay et al. 1990; Müller et al. 1994).

A database comprising representative petrological and geochemical data from the potassic and shoshonitic intrusive rocks from these mineralized localities was com-

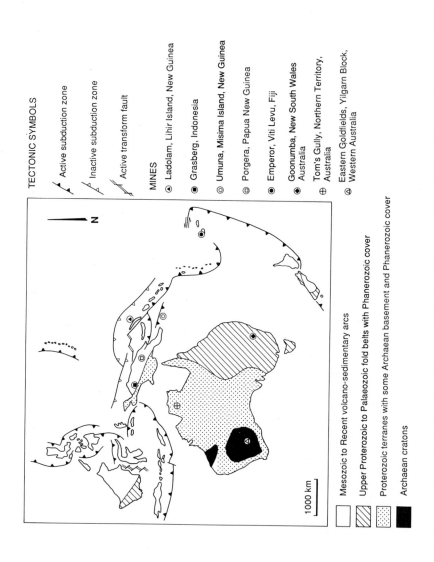

TECTONIC SYMBOLS

Active subduction zone

Inactive subduction zone

Active transform fault

MINES

④ Ladolam, Lihir Island, New Guinea

◉ Grasberg, Indonesia

◎ Umuna, Misima Island, New Guinea

◉ Porgera, Papua New Guinea

◉ Emperor, Viti Levu, Fiji

◉ Goonumba, New South Wales
   Australia

⊕ Tom's Gully, Northern Territory,
   Australia

◐ Eastern Goldfields, Yilgarn Block,
   Western Australia

N

1000 km

☐ Mesozoic to Recent volcano-sedimentary arcs

▨ Upper Proterozoic to Palaeozoic fold belts with Phanerozoic cover

▦ Proterozoic terranes with some Archaean basement and Phanerozoic cover

■ Archaean cratons

Fig. 6.1.   Overview in terms of gross tectonic setting of some major gold and base-metal deposits hosted by potassic igneous rocks in the Southwest Pacific area. Modified after Müller and Groves (1993). See Figure 1.1 for a more complete overview of the entire Pacific Rim.

Table 6.1. Some representative major gold and base-metal deposits associated with potassic igneous rocks through time.

Age	Deposit/ mineral field	Location	Mineralization	Host rocks	Critical geochemistry	Alteration	Tectonic setting	References
Pleistocene	Ladolam	Lihir Island, Papua New Guinea	Epithermal gold	Latites, trachybasalts	High LILE Low LREE Low HFSE	Potassic and propylitic	Late oceanic arc	Moyle et al. (1990) Wallace et al. (1983)
Pliocene	Emperor	Viti Levu, Fiji	Epithermal gold	Monzonites, trachyandesites	High LILE Low LREE Low HFSE	Potassic and acid sulphate	Late oceanic-arc	Ahmad (1979) Ahmad and Walshe (1990) Anderson and Eaton (1990) Kwak (1990) Setterfield (1991) Setterfield et al. (1991, 1992)
Pliocene	Grasberg	Indonesia	Porphyry copper-gold	Diorites	High LILE Low LREE Low HFSE	Potassic and propylitic	Postcollisional arc	MacDonald and Arnold (1994)
Pliocene	Misima	Papua New Guinea	Epithermal gold	Shoshonitic lamprophyres	High LILE Mod. LREE High Nb	Phyllic and propylitic	Postcollisional arc	Appleby et al. (1995)
Miocene	Porgera	Papua New Guinea	Epithermal gold	Feldspar porphyry dykes, trachybasalts	High LILE Low LREE High Nb	No potassic alteration	Postcollisional arc	Handley and Henry (1990) Richards (1990a, 1990b, 1992) Richards et al. (1990, 1991)
Miocene	El Indio	Chile	Epithermal gold	Monzodiorites	High LILE Low HFSE	Argillic	Continental arc	Jannas et al. (1990)
Miocene	Bajo de la Alumbrera	Argentina	Porphyry copper-gold	Dacites	High LILE Mod. LREE Low HFSE	Potassic	Continental arc	Müller and Forrestal (1998)
Ordovician	Goonumbla	New South Wales, Australia	Porphyry copper-gold	Monzonites, trachytes, trachyandesites	High LILE Low LREE Low HFSE	Potassic and propylitic	Late oceanic arc	Heithersay et al. (1990) Müller et al. (1994)
Proterozoic	Tom's Gully	Northern Territory, Australia	Mesothermal gold	Granites, syenites, shoshonitic lamprophyres	High LILE High LREE High HFSE	Secondary carbonate	Within plate	Sheppard (1992) Sheppard and Taylor (1992)
Archaean	Superior Province	Canada	Mesothermal gold	Shoshonitic lamprophyres	High LILE Low LREE Low HFSE	Secondary carbonate	Postcollisional arc	Wyman (1990) Wyman and Kerrich (1988, 1989a, 1989b)
Archaean	Yilgarn Block	Western Australia	Mesothermal gold	Shoshonitic lamprophyres	High LILE Low LREE Low HFSE	Secondary carbonate	Postcollisional arc	Perring et al. (1989a) Rock (1991) Rock et al. (1989) Taylor et al. (1994)

Table 6.2. Whole-rock data sources for samples in database GOLD1. The number in square brackets refers to the number of analyses from that reference in the database. From Müller and Groves (1993).

1. Continental arcs	2. Postcollisional arcs	3a. Late oceanic arcs	4. Within-plate settings
El Indio, Chile Tschischow (1989) [5]	*Misima, Papua New Guinea* K. Appleby (pers. comm., 1996) [5]	*Viti Levu, Fiji* Gill (1970) [10] Setterfield (1991) [27]	*Mount Bundey, Northern Territory, Australia* Sheppard and Taylor (1992) [6]
Bajo de la Alumbrera, Argentina Müller and Forrestal (1998) [4]	*Porgera, Papua New Guinea* Richards (1990a) [6] Richards (1990b) [13] Richards et al. (1990) [4]	*Goonumbla, New South Wales, Australia* Müller et al. (1994) [3]	
Bingham, Utah, USA Waite et al. (1997) [3]		*Lihir Island, Papua New Guinea* Wallace et al. (1983) [16] Kennecott Exploration Co. (pers. comm., 1992) [2]	
	Superior Province, Canada Wyman (1990) [20] Wyman and Kerrich (1989a) [31]	*Dinkidi; Didipio, Phillipines* Wolfe (1999) [3]	
	Yilgarn Block, Western Australia Taylor et al. (1994) [18] The University of Western Australia (unpubl. data) [15]		

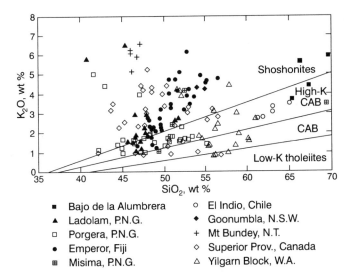

Fig. 6.2. K_2O versus SiO_2 plot (after Peccerillo and Taylor 1976a) showing data from database GOLD1. CAB = calc-alkaline basalts. Adapted from Müller and Groves (1993).

piled (GOLD1), and the sources are listed in Tables 6.1 and 6.2. Available samples were filtered and geochemical discrimination diagrams developed. The data were plotted on the K_2O versus SiO_2 biaxial diagram of Peccerillo and Taylor (1976a) to illustrate the potassic affinities of the igneous rocks (Fig. 6.2). Rocks from the gold deposits at Emperor, Tom's Gully (Mount Bundey), Ladolam, and Porgera generally show the highest K_2O contents (Fig. 6.2). Mount Bundey is a potassic pluton close to the Tom's Gully gold deposit and both are associated with potassic lamprophyres.

Nearly all of the potassic igneous rocks and shoshonites from mineralized environments investigated in this study are arc related (cf. Müller and Groves 1993), as illustrated by their geological setting and confirmed by their position on the TiO_2 versus Al_2O_3 and Y versus Zr biaxial plots (Figs 6.3, 6.4). However, spatial associations between some potassic lamprophyres and mesothermal gold mineralization, for example at the Tom's Gully deposit in the Proterozoic Pine Creek Geosyncline, Northern Territory, Australia (see Sect. 7.3), occur in within-plate tectonic settings as illustrated in Figures 6.3 and 6.4.

6.2 Erection of Database GOLD1

A new database (GOLD1) has been constructed so that geochemical discrimination diagrams — based on the filtered database SHOSH2 —can be applied to mineralized tectonic settings. The database comprises only geochemical data for potassic

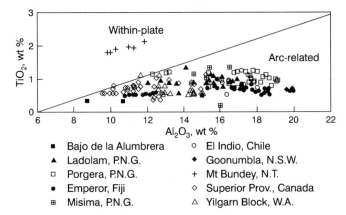

Fig. 6.3.   TiO$_2$ versus Al$_2$O$_3$ diagram (see Chap. 3) showing discriminant fields and the position of shoshonitic and potassic igneous rocks associated with gold-copper mineralization. Mount Bundey is the intrusion adjacent to the Tom's Gully gold deposit, Northern Territory, Australia. Adapted from Müller and Groves (1993).

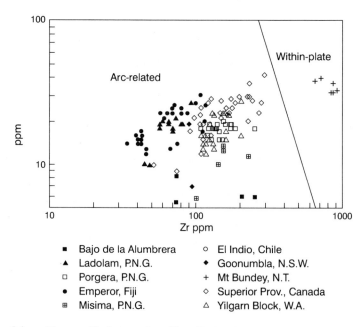

Fig. 6.4.   Y versus Zr diagram (see Chap. 3) showing discriminant fields and the position of shoshonitic and potassic igneous rocks associated with gold-copper mineralization. Adapted from Müller and Groves (1993).

igneous rocks from mineralized localities such as the Emperor, Ladolam, and Porgera gold deposits, and major porphyry-copper districts such as the Chilean Andes and the Goonumbla area, Lachlan Fold Belt, as listed in Table 6.2. The data were taken from the literature or derived from analyses undertaken for this study. Using the procedure given in Chapter 3, samples were filtered and discrimination diagrams developed. Although several samples from highly mineralized settings (e.g. Porgera) have high LOI values (i.e. > 5 wt %), and slightly higher CaO (up to 2.4 wt %) and Na_2O (up to 5.3 wt %) contents than recommended for the use of the discrimination diagrams, they were left in the database GOLD1.

6.3   Late Oceanic Arc Associations

Generally, the samples from those mineralized arcs described below are characterized by the lowest Zr concentrations (< 110 ppm, see Fig. 6.4) and hence are typical of potassic volcanic rocks from oceanic arc settings (Chap. 3). Figure 6.5 clearly

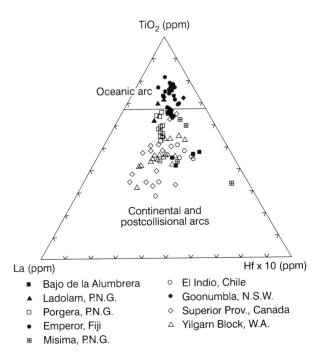

Fig. 6.5. $(TiO_2/100)$-La-$(Hf \times 10)$ triangular diagram (see Chap. 3) discriminating between samples from oceanic arcs (black symbols) and those from continental or postcollisional arc settings (white symbols). The symbols are the same as in Figures 6.2–6.4. Adapted from Müller and Groves (1993).

discriminates them from high-K rocks typical of continental and postcollisional arc settings, which are also characterized by higher Zr contents (up to 300 ppm; see Fig. 6.4). As demonstrated in Figure 6.6 — the $(TiO_2/10)$-$(10 \cdot La)$-$(P_2O_5/10)$ triangular plot of Chapter 3 — all mineralized potassic island-arc rocks discussed here are from late oceanic arcs, with the implication that initial oceanic arcs may be barren of gold or base-metal mineralization in association with high-K igneous rocks.

6.3.1   Ladolam Gold Deposit, Lihir Island, Papua New Guinea

Introduction. The Southwest Pacific hosts some of the world's premier gold and gold-copper deposits (Andrews 1995). One example is Ladolam, on Lihir Island, which is hosted by Quaternary shoshonitic rocks (Wallace et al. 1983; Moyle et al. 1990). Lihir Island is in the New Ireland Province of Papua New Guinea, and has an area of about 192 km². The Ladolam gold deposit is located on the east coast of Lihir Island (Fig. 6.7) and was discovered in 1982 by the Kennecott Explorations and Niugini Mining Joint Venture (Hoogvliet 1993).

Regional Geology. Lihir Island is one of four volcanic island groups which form a chain parallel to the New Ireland coast line. The other islands are Tabar to the north-

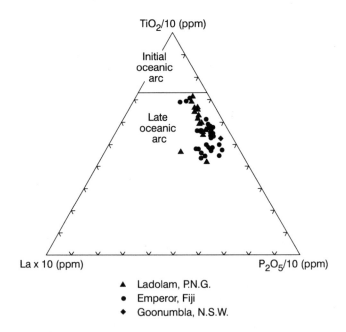

Fig. 6.6.   Samples from oceanic arcs (Lihir Island, Viti Levu, Goonumbla) plotted on the $(TiO_2/10)$-$(Lax10)$-$(P_2O_5/10)$ triangular diagram (see Chap. 3) suggesting their genesis in a late oceanic arc. From Müller and Groves (1993).

west and Tanga and Feni to the southeast (Wallace et al. 1983). The islands are composed of Pliocene-Pleistocene subaerial volcanic rocks. Wallace et al. (1983) interpret the shoshonites to be derived from subduction-modified mantle during splitting of the Melanesian Arc and opening of the Manus Basin southwest of New Ireland.

Lihir Island consists of lavas, volcanic breccias, pyroclastic rocks, and derivative epiclastic rocks (Moyle et al. 1990). The Ladolam gold deposit is situated within the elliptical caldera of a Quaternary stratovolcano with dimensions of 5.5 by 3.5 km.

Nature of Epithermal Gold Mineralization. There are four mineralized areas at Ladolam — Minifie, Lienetz, Coastal and Kapit — which all occur within the caldera centre (Hoogvliet 1993). The mineralization consists of sulphidic breccias and gold-bearing pyrite veining, ranging from microscopic stringers to massive veins within the brecciated volcanic rocks and intrusions (Moyle et al. 1990). Gold mineralization is believed to be of a magmatic origin, and the magmas forming the island arc are intrinsically enriched in gold (Moyle et al. 1990). Gold, which was probably transported by near-neutral chloride fluids (Moyle et al. 1990), is primarily contained in refractory pyrite ore between sea level and 200 m below (H. Hoogvliet, pers. comm. 1993).

The rocks are strongly affected by argillic and phyllic alteration (Hoogvliet 1993).

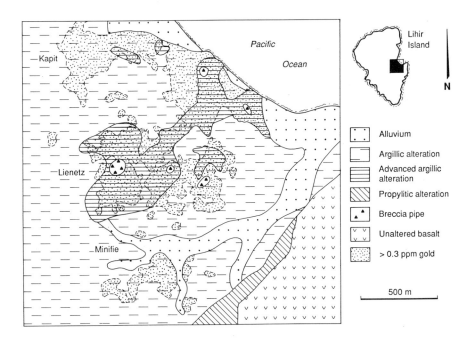

Fig. 6.7.   Alteration map of the Ladolam gold deposit, Lihir Island, Papua New Guinea. Modified after Dimock et al. (1993).

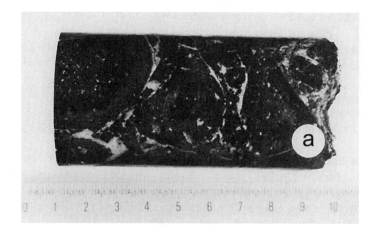

Fig. 6.8.   Hand specimens of autoclastic breccias from the Ladolam gold deposit, Papua New Guinea. (a) 119429. (b) 119430. (c) 119431. Sample numbers refer to specimens held in the Museum of the Department of Geology and Geophysics, The University of Western Australia.

Argillic alteration is defined by secondary smectite, kaolinite, and illite, whereas phyllic alteration is characterized by illite, which is commonly associated with secondary orthoclase and silica. High-grade epithermal gold mineralization is superimposed mainly on potassically altered volcanic rocks, and areas of propylitic alteration typically contain only low gold grades (Moyle et al. 1990).

The combined proven, probable, and possible geological resources of Ladolam are about 425 million tonnes of ore averaging 2.8 ppm Au at a cut-off of 1.5 ppm; mineable sulphide ore reserves are quoted as 89.3 million tonnes averaging 4.7 ppm Au at a cut-off of 2 ppm (Hoogvliet 1993). The dominant primary opaque minerals at Ladolam are pyrite and marcasite with minor magnetite, rutile, galena, and sphalerite, and rare gold-silver tellurides (Moyle et al. 1990).

Lihir Island is still geothermally active, with evidence for active gold deposition (Moyle et al. 1990), as already reported from Feni Island (Wallace et al. 1983).

Petrology and Geochemistry of the Shoshonitic Host Rocks. The host rocks consist of porphyritic trachybasaltic, trachyandesitic and latitic lavas, and pyroclastic rocks with high K contents (Wallace et al. 1983) which, in some places, are cut by monzonite intrusions. Several andesitic to latitic porphyry stocks and dykes have been intersected in drill holes (Moyle et al. 1990). Volcanic rocks dominating the upper parts of the deposit consist of lavas, tuffs, and subvolcanic breccias (Fig. 6.8). They are porphyritic with phenocrysts of plagioclase, augite, and minor biotite and hornblende in a feldspathic groundmass (Moyle et al. 1990). The monzonitic intrusions are fine to medium grained, quartz poor, and contain plagioclase, orthoclase, augite, and biotite. Lithostatic unloading during caldera formation resulted in widespread hydraulic brecciation of the host rocks that was healed by anhydrite and calcite veining (Moyle et al. 1990).

All intrusive and volcanic high-K rocks fall into the field of late oceanic-arc magmas, as shown in Figures 6.5 and 6.6. The rocks have relatively evolved compositions (Table 6.3), with mg# of ~ 55 and relatively low concentrations of the mantle-compatible trace-elements (e.g. < 95 ppm Cr, < 20 ppm Ni). Their very high LILE (e.g. up to 128 ppm Rb, 1440 ppm Sr), moderate LREE (e.g. ~ 16 ppm La, ~ 40 ppm Ce), and very low HFSE (e.g. < 1.0 wt % TiO_2, < 75 ppm Zr, ~ 2 ppm Nb, < 2 ppm Hf) contents are typical for potassic igneous rocks derived in a late oceanic-arc setting (Müller et al. 1992b). The potassic rocks from the Ladolam gold deposit also have relatively high halogen contents (see Sect. 8.3.2). This observation is consistent with the recent discovery of submarine, gold-rich hydrothermal vents associated with volatile-rich alkaline rocks in the sea only 25 km south of Lihir Island (Herzig and Hannington 1995).

Table 6.3.   Major- and trace-element analyses of potassic igneous rocks from Lihir Island, Papua New Guinea. Major elements are in wt %, and trace elements are in ppm. Fe_2O_3 (tot) = total iron calculated as ferric oxide. From Wallace et al. (1983).

Province/deposit:	Lihir Island	Lihir Island
Location:	Papua New Guinea	Papua New Guinea
Rock type:	Basalt	Basalt
Tectonic setting:	Late oceanic arc	Late oceanic arc
Reference:	Wallace et al. (1983)	Wallace et al. (1983)
SiO_2	46.70	47.90
TiO_2	1.04	0.87
Al_2O_3	16.10	15.20
Fe_2O_3 (tot)	10.90	10.70
MnO	0.21	0.22
MgO	5.85	5.85
CaO	12.10	12.00
Na_2O	3.05	3.40
K_2O	1.87	2.90
P_2O_5	0.54	0.53
LOI	1.89	0.50
Total	100.25	100.07
mg#	55	55
V	251	275
Cr	24	94
Ni	14	20
Rb	128	59
Sr	1430	1440
Y	21	19
Zr	74	67
Nb	2	2
Ba	265	270
La	15	16
Ce	37	40
Hf	1.8	1.6

6.3.2 Emperor Gold Deposit, Viti Levu, Fiji

Introduction. Epithermal gold mineralization in a late oceanic arc setting occurs at the Emperor mine, Viti Levu, Fiji (Fig. 6.9). Mineralization is hosted by Pliocene shoshonitic volcanic rocks (Ahmad 1979; Anderson and Eaton 1990; Setterfield 1991; R. Jones, pers. comm. 1993).

Regional Geology. Fiji is situated at the boundary between the Indo-Australian and Pacific plates, midway between the west-dipping Tonga-Kermadec Trench and the

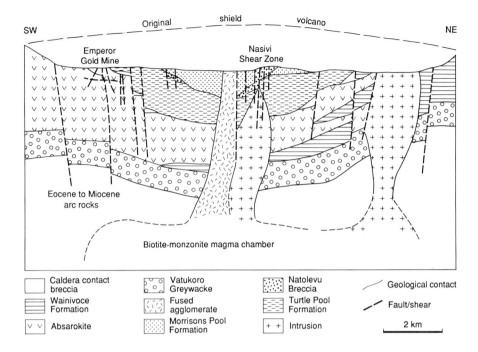

Fig. 6.9.   Geological overview of the Emperor gold deposit, Fiji. Modified after Setterfield et al. (1991).

east-dipping Vanuatu Trench (Gill and Whelan 1989). Magmatism in Fiji is largely derived from subduction along the now-inactive Vitiaz Trench (Setterfield et al. 1992).

Viti Levu was formed during three periods of volcanism, with the igneous rock series erupted, in order, from tholeiites, to calc-alkaline, and finally shoshonitic rocks (Gill 1970; Setterfield 1991).

The Emperor deposit supports the country's only operating gold mine and is located in northern Viti Levu at the caldera margin of a Tertiary shoshonitic shield volcano (Anderson and Eaton 1990). The deposit is situated at the intersection of a major northeast-trending lineament with a lesser northwest-trending shear zone (Setterfield et al. 1991).

Nature of Epithermal Gold Mineralization. Mineralization in Fiji includes various types of volcanogenic massive-sulphide (VMS) deposits, disseminated porphyry copper-gold deposits and epithermal gold-tellurium-silver deposits, most of which are associated with high-K calc-alkaline and shoshonitic rocks (Colley and Greenbaum 1980). The Emperor mine has produced about 125 tonnes of gold to date (Setterfield et al. 1992). Gold mineralization occurs as epithermal quartz-sericite±carbonate veining containing gold-silver tellurides and auriferous pyrite at the western caldera margin

Table 6.4.   Major- and trace-element analyses of shoshonitic rocks from the Emperor gold deposit, Fiji. Major elements are in wt %, and trace elements are in ppm. Fe_2O_3 (tot) = total iron calculated as ferric oxide. From Setterfield (1991).

Province/deposit:	Emperor	Emperor
Location:	Fiji	Fiji
Rock type:	Basalt	Basalt
Tectonic setting:	Late oceanic arc	Late oceanic arc
Reference:	Setterfield (1991)	Setterfield (1991)
SiO_2	51.29	52.73
TiO_2	0.60	0.54
Al_2O_3	17.87	19.71
Fe_2O_3 (tot)	7.92	6.29
MnO	0.17	0.15
MgO	3.15	1.98
CaO	6.51	5.84
Na_2O	3.69	3.00
K_2O	4.70	6.13
P_2O_5	0.53	0.48
LOI	3.18	2.73
Total	99.61	99.58
mg#	48	42
V	198	146
Cr	11	3
Ni	6	2
Rb	142	105
Sr	1571	1812
Y	21	23
Zr	85	87
Nb	2	4
Ba	1209	1145
La	23	18
Ce	39	34
Hf	2	2.1

(Setterfield et al. 1992), and is accompanied by propylitic alteration (Setterfield 1991) and, in places, potassic alteration of the host rocks (Ahmad and Walshe 1990; Anderson and Eaton 1990). ^{40}Ar-^{39}Ar dating indicates that epithermal gold mineralization at Emperor was formed at 3.71 ± 0.13 Ma, whereas shoshonite emplacement and the less important porphyry-style mineralization were earlier at ca. 4.3 Ma (Setterfield 1991). Stable isotope and fluid inclusion studies suggest that both gold mineralization and potassic alteration were produced by ascending magmatic fluids mixed with heated meteoric waters (Ahmad and Walshe 1990; Anderson and Eaton 1990; Kwak 1990), with the magmatic source being a high-K monzonite stock at depth (Ahmad 1979).

Petrology and Geochemistry of the Shoshonitic Host Rocks. The basaltic host rocks at Emperor have a shoshonitic geochemistry and consist of plagioclase, augite, and olivine phenocrysts, with minor biotite and hornblende, in a fine-grained feldspathic groundmass (Gill 1970; Ahmad and Walshe 1990; Setterfield et al. 1991). Magma evolution was controlled solely by fractional crystallization of titanomagnetite, olivine, clinopyroxene, and plagioclase (Setterfield 1991). All volcanic rocks plot in the late oceanic-arc magma series on Figures 6.5 and 6.6.

Fractional crystallization resulted in low mg# (< 50), and low concentrations of mantle-compatible trace-elements (e.g. ~ 10 ppm Cr, ~ 5 ppm Ni). The rocks, which are highly potassic (up to 6.1 wt % K_2O), also contain high concentrations of Al_2O_3 (up to 19.7 wt %) and Na_2O (up to 3.6 wt %) (Setterfield 1991). Their trace-element composition (Table 6.4) is characterized by very high LILE (e.g. up to 142 ppm Rb, up to 1800 ppm Sr, up to 1200 ppm Ba), moderate LREE (e.g. ~ 20 ppm La, ~ 40 ppm Ce), and very low HFSE (e.g. < 0.6 wt % TiO_2, < 90 ppm Zr, < 4 ppm Nb, ~ 2 ppm Hf) contents, which are typical for shoshonitic rocks derived in a late oceanic-arc setting (Müller et al. 1992b).

6.3.3 Dinkidi Copper-Gold Deposit, Didipio, Philippines

Introduction. The Dinkidi porphyry copper-gold deposit is located in the Didipio area (Fig. 6.10) about 200 km north of Manila, in the remote Sierra Madre Mountains of the Nueva Vizcaya Province of northern Luzon, Philippines (Haggman 1997a). Access to the site is via logging roads from Cabarroguis (Garrett 1996). Dinkidi was discovered by Climax Arimco Corporation in 1988.

The region was mined for alluvial gold for many years, as Dinkidi forms a razorback ridge protruding from an alluvial valley floor (Haggman 1997a). On the ridge there is narrow quartz-stockwork veining and zones with argillic alteration, which overprints monzodioritic to monzonitic intrusions (Haggman 1997a). Systematic exploration of the ridge surface revealed abundant copper-oxide minerals. A 300 m by 500 m-wide anomaly was identified by a subsequent IP survey (Fig. 6.10), and diamond drilling intersected porphyry copper-gold mineralization (Haggman 1997a). Detailed geological mapping of the area suggests that Dinkidi forms part of a monzonitic caldera complex measuring about 8 km in diameter (Haggman 1997b).

The Dinkidi deposit consists of an indicated resource of 120 million tonnes at 1.2 ppm Au and 0.5 wt % Cu (Wolfe et al. 1998), including a high-grade zone of 7.9 million tonnes at 6 ppm Au equivalent (Haggman 1997b). Oxidized ore represents only three percent of the total resource.

Regional Geology. The Didipio area consists of late-Cretaceous to mid-Miocene volcanosedimentary and volcaniclastic rocks that were derived in an oceanic (island) arc setting (Haggman 1997a).

Dinkidi is though to be related to the change from Miocene westward subduction along the northern Luzon Trench to eastward subduction along the Manila Trench (Solomon 1990; Corbett and Leach 1998). Dinkidi occurs in a back arc basin that

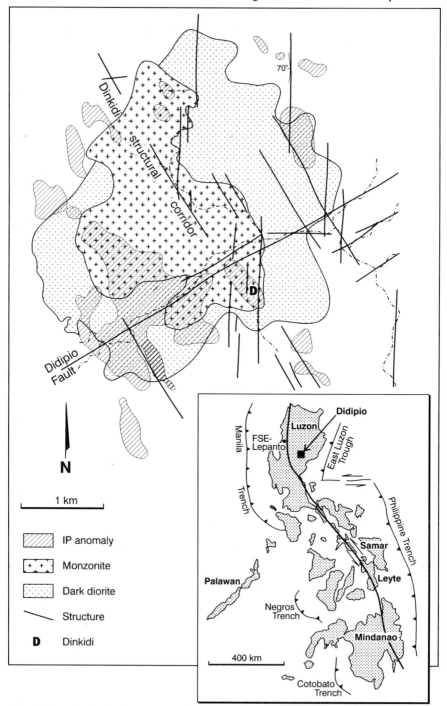

Fig. 6.10. Geological overview of the Dinkidi copper-gold deposit, Didipio, Philippines. Modified after Haggman (1997b) and Wolfe et al. (1998).

trends north-south along the Cagayan Valley (Haggman 1997b). The Didipio Igneous Complex represents several potassic igneous intrusions that are localized by the intersection of east-northeast-trending transfer structures with north-northwest fractures parallel to the Philippine Fault (Haggman 1997b).

Stratigraphically, the Didipio area is formed by a pre-Tertiary basement complex of tonalites and schists, which are covered by Eocene andesitic lavas and basaltic tuffs interlayered with sedimentary rocks (Haggman 1997b). This Eocene Caraballo Group is unconformably overlain by the Late Oligocene Mamparang Formation, which comprises calc-alkaline andesitic lavas and volcaniclastic rocks that have been intruded by the early-Miocene Didipio Igneous Complex (Wolfe et al. 1998).

The Didipio Igneous Complex consists of a series of early clinopyroxene cumulates, diorites, and monzodiorites that were intruded by a large, late-stage monzonite pluton (Wolfe et al. 1998). The Dinkidi deposit is hosted by a large, 800 m by 200 m wide, elongate monzonite stock at the southern margin of this composite monzonite pluton. The stock predates and hosts the bulk of the porphyry copper-gold mineralization at Dinkidi (Wolfe 1999).

Nature of Porphyry Copper-Gold Mineralization. The mineralized monzonite at Dinkidi forms a steeply dipping tabular body, measuring about 500 m by 200 m, with mineralization being open at a depth of 800 m (Haggman 1997a). The monzonite was intruded by a thin, highly mineralized, variably textured magnetite- and clinopyroxene-phyric monzonitic pegmatite, which is interpreted to represent a volatile-rich late-stage felsic differentiate from the monzonitic magma (Wolfe et al. 1998). Recent studies (R. Wolfe, written comm. 1999) imply that these intrusions are co-magmatic.

Alteration textures, mineralogy, and intensity vary broadly throughout the deposit (Garrett 1996). The deep core of the deposit is characterized by an intense whitish clay-carbonate-muscovite-silica-sericite alteration, which overprinted the monzonite porphyry. The alteration is accompanied by weak disseminations of magnetite-pyrite-chalcopyrite and, more rarely, fine veinlets of quartz-pyrite-chalcopyrite (Garrett 1996). This alteration type grades vertically upward into a potassic alteration assemblage (Garrett 1996). Within the potassic alteration zone, the plagioclase phenocrysts are typically altered to white sericite-clay-carbonate assemblages, whereas the groundmass fraction is flooded by pinkish-grey orthoclase (Garrett 1996). The potassic alteration at Dinkidi is dominated by secondary orthoclase (Corbett and Leach 1998). The less abundant clinopyroxene and hornblende phenocrysts of the monzonite intrusion are weakly altered to chlorite-magnetite assemblages.

The principal copper-sulphides are chalcopyrite and, to a lesser extent, bornite, and they commonly occur as fine disseminations and within quartz stockwork veins (Haggman 1997b). Bornite occurs as alteration rims around and within chalcopyrite grains (Haggman 1997b). Native gold occurs as inclusions within chalcopyrite and, less commonly, within bornite. Native gold and electrum are rare inclusions in silicates (Garrett 1996).

High-grade porphyry copper-gold mineralization (> 3 ppm Au, > 1 ppm Cu) is either associated with potassic alteration or occurs within the silica-carbonate-py-

Table 6.5.   Major- and trace-element analyses of potassic igneous rocks from the Dinkidi copper-gold deposit, Didipio, Philippines. Major elements are in wt %, and trace elements are in ppm. Fe_2O_3 (tot) = total iron calculated as ferric oxide. Data from Wolfe (1991).

Province/deposit:	Dinkidi	Dinkidi	Dinkidi
Location:	Philippines	Philippines	Philippines
Rock type:	Monzodiorite	Monzonite	Quartz monzonite
Tectonic setting:	Late oceanic arc	Late oceanic arc	Late oceanic arc
Reference:	Wolfe (1999)	Wolfe (1999)	Wolfe (1999)
SiO_2	51.81	57.55	61.40
TiO_2	0.65	0.45	0.37
Al_2O_3	18.66	18.29	19.19
Fe_2O_3 (tot)	8.28	5.44	4.11
MnO	0.19	0.13	0.07
MgO	3.86	2.09	1.40
CaO	9.54	5.65	2.97
Na_2O	3.75	4.87	5.58
K_2O	2.57	4.13	4.38
P_2O_5	0.46	0.37	0.23
LOI	0.23	0.76	0.99
Total	100.01	99.73	100.69
mg#	52	47	44
Cr	6	3	2
Ni	7	3	1
Rb	49	70	79
Sr	1005	872	864
Y	20	18	13
Zr	66	100	72
Nb	1.5	1.9	1.7
Ba	502	484	1100
La	11	15	9
Ce	24	29	20
Yb	2.0	1.8	1.3

rite-chalcopyrite zone surrounding the core of the deposit (Haggman 1997b). Outside the potassic alteration zone, the grades are lower (Garrett 1996).

In places, hydrothermal brecciation overprints the potassic alteration zone. These hydrothermal breccias generally contain the highest copper-gold grades due to remobilization of these metals and their concentration within the breccia matrix (Garrett 1996). The hydrothermal brecciation is consistent with the suspected high volatile contents of the parental melts.

The porphyry copper-gold mineralization is interpreted to be genetically related to the late-stage release of volatiles and monzonitic magma from the magma chamber beneath the caldera. The fractal nature of the composite monzonite stock is consistent with the release of increasingly more-evolved fractions from a larger magma chamber at depth.

Dinkidi has several similarities with the Goonumbla porphyry copper-gold deposit in New South Wales, Australia (Sect. 6.3.4). Both deposits are hosted by porphyritic monzonite intrusions that represent the late-stage melts derived from a magma chamber underlying a caldera setting. At both deposits, the monzonites have high-K compositions and are interpreted to have formed in late oceanic arcs.

Petrography and Geochemistry of the Potassic Host Rocks. The mineralized monzonite porphyry is a pinkish-grey, medium-grained, intrusive rock consisting of phenocrysts of plagioclase, clinopyroxene, biotite, and hornblende, which are set in a feldspathic groundmass (Garrett 1996). The monzonite also contains apatite microphenocrysts. The rock has a vughy texture, but is intensely and pervasively altered. Its groundmass is flooded by secondary orthoclase (Garrett 1996). The vughy relic texture is interpreted to reflect the volatile-rich nature of the parental melts, a common feature of most porphyry copper-gold deposits hosted by potassic igneous rocks.

Table 6.5 shows representative whole-rock analyses of the potassic igneous rocks from the Didipio Igneous Complex (Wolfe 1999). The rocks are characterized by enriched LILE concentrations (e.g. up to 4.38 wt % K_2O; up to 79 ppm Rb, 1005 ppm Sr, 1100 ppm Ba), low LREE contents (e.g. < 15 ppm La, < 29 ppm Ce), and very low HFSE abundances (e.g. TiO_2 < 0.65 wt %, < 100 pm Zr, < 2 ppm Nb). Fractional crystallization resulted in low mg# (< 52), and very low concentrations of mantle-compatible trace-elements (e.g. < 6 ppm Cr, < 7 ppm Ni). The rocks have elevated P_2O_5 concentrations (up to 0.46 wt %), which is consistent with the presence of apatite microphenocrysts.

The samples have not been plotted on geochemical discriminant diagrams due to a lack of Hf analyses. However, based on their geochemical compositions, the rocks are interpreted to be derived in a late oceanic-arc setting (Müller et al. 1992b).

6.3.4 Goonumbla Copper-Gold Deposit, New South Wales, Australia

Introduction. An example of ancient mineralization hosted by shoshonitic rocks in a late oceanic-arc setting is the Ordovician Goonumbla porphyry copper-gold deposit in New South Wales, Australia (Heithersay et al. 1990; Müller et al. 1994; Heithersay and Walshe 1995; Hooper et al. 1996). It is described in some detail here because many of the primary data were collected specifically for this study (cf. Müller 1993). The Goonumbla igneous complex is situated within a collapsed caldera and hosts the world-class Northparkes copper-gold mine. This igneous suite forms one of several mineralized Ordovician shoshonitic centres in the Lachlan Fold Belt (Thompson et al. 1986; Perkins et al. 1992; Wyborn 1992). Eleven centres of volcanic-hosted mineralization can be distinguished (Hooper et al. 1996). The largest deposit is Endeavour 26 North (see E26N on Fig. 6.11), with estimated ore resources of 166 million tonnes at 0.74 wt % Cu, 0.12 ppm Au, and 1.7 ppm Ag (Heithersay 1986). The Endeavour 26 North deposit is currently being mined at Northparkes.

Regional Geology. Many igneous provinces of Ordovician age are known from south-eastern Australia (Horton 1978; Powell 1984; Thompson et al. 1986; Perkins et al. 1992; Wyborn 1992; Müller et al. 1993a). The Lachlan Fold Belt in New South Wales is divided into several north-south-trending synclinorial and anticlinorial tectonic zones composed of Palaeozoic igneous and sedimentary rocks (Scheibner 1972). The Goonumbla district is situated in the northeastern part of the Bogan Gate synclinorial zone (Scheibner 1974). This north-trending trough forms a broad tectonic boundary within the Lachlan Fold Belt, separating two Proterozoic terranes to the west and east of the Parkes area — the Wagga Metamorphic Belt and the Kosciusko Terrane, respectively — and comprises a zone of Palaeozoic sedimentary rocks with associated late Ordovician to early Silurian igneous rocks (Jones 1985; Perkins et al. 1992). In the Parkes area, a sequence of latitic lavas, interlayered flows and pyroclastic units with minor volcaniclastic rocks and limestones comprises the Goonumbla igneous complex (Jones 1985). The rocks are generally unmetamorphosed, but are commonly gently folded with relatively shallow dips (Heithersay et al. 1990). Com-

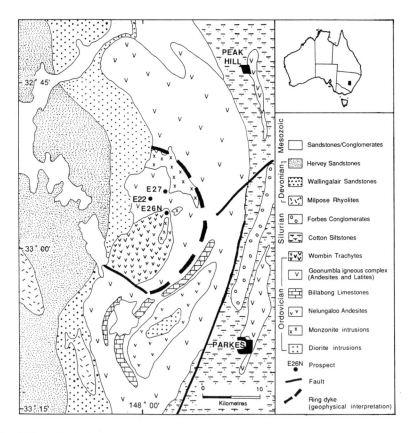

Fig. 6.11. Geological overview of the Goonumbla igneous complex, New South Wales, Australia Modified after Heithersay et al. (1990).

positions of the igneous rocks, which are locally intruded by monzonites, range from andesitic to trachytic, and they exhibit typical shoshonitic geochemistry (Joplin et al. 1972; Morrison 1980). As discussed by Müller et al. (1994), the Goonumbla igneous rocks are probably related to a former subduction event.

Nature of Porphyry Copper-Gold Mineralization. Within the Goonumbla district, eleven centres of mineralization are located within a circular feature which is some 22 km in diameter (Fig. 6.11). It is partly bounded by a monzonitic ring dyke, and is interpreted to be a collapsed caldera formed as a result of regional extension (Jones 1985). The Goonumbla igneous rocks consist mainly of andesites, latites, and the slightly younger Wombin trachytes locally intruded by monzonite stocks and minor basaltic dykes. The igneous rocks represent a cogenetic suite with shoshonitic geochemistry as defined by Morrison (1980).

The mineralized igneous host rocks comprise a repetitive sequence of andesitic lavas, latites and trachytes, with associated epiclastic rocks. Porphyry copper-gold mineralization in the Goonumbla district is generally associated with relatively small pipe-like intrusive bodies of quartz-rich monzonite (e.g. Fig. 6.12). These monzonite stocks can have diameters up to 100 m and vertical extensions up to 900 m (Perkins et al. 1992), and were probably formed as late-stage differentiates of andesitic parental melts. These late-stage differentiates of quartz-monzonite form fractal, finger-like intrusions which commonly have crowded porphyritic centres that grade outward, through zones with mosaic textures, into sparsely porphyritic margins (G. Morrison, pers. comm. 1996). The highest volatile concentrations appear to be in the mosaic-textured zones.

Primary copper and zinc mineralization consists of disseminated and vein sulphides, notably chalcopyrite, bornite, chalcocite, sphalerite and minor pyrite, and is generally associated with quartz stockwork veining which occurs both within the intrusive bodies and the surrounding volcanic host rocks (Heithersay and Walshe 1995). Sulphide mineralization is commonly accompanied by disseminated grains of haematite and magnetite. Pervasive haematite-sericite alteration occurs peripheral to the mineralization.

Native gold occurs mainly as minute grains within silicates of the host rock, and more rarely as fine inclusions in the sulphides (Jones 1985). The highest gold values are present in the potassic alteration zone (see below), and are closely associated with chalcopyrite and bornite mineralization.

No large pervasive alteration zones that typify many other porphyry copper deposits (e.g. Bajo de la Alumbrera, Argentina; see Gonzalez 1975) are present in the Goonumbla igneous complex (Heithersay et al. 1990). However, mineralization is generally associated with potassic alteration zones, characterized by either pervasive secondary biotite flakes or hydrothermal sericite or orthoclase veins. Pink hydrothermal orthoclase also forms dense granular patches. Secondary biotite is most commonly developed as dark pervasive zones within the more mafic andesites.

Regional zones of propylitic alteration in the Goonumbla district are mainly characterized by irregular secondary blebs of epidote, chlorite, and carbonate in igneous rocks. Epidote also forms fine-grained aggregates which have replaced primary mica

Table 6.6.   Major- and trace-element analyses of shoshonitic rocks from the Goonumbla porphyry copper-gold deposit, New South Wales, Australia. Major elements are in wt %, and trace elements are in ppm. Samples were derived from diamond drill cores of the E26N or E27 prospects. Sample numbers refer to specimens held in the Museum of the Department of Geology and Geophysics, The University of Western Australia. Data from Müller et al. (1994).

Sample no.:	119082	119083	119084	119085
Origin:	E27	E27	E27	E26N
Rock type:	Basalt	Andesite	Andesite	Andesite
SiO_2	46.8	54	54.9	56.7
TiO_2	0.57	0.67	0.67	0.64
Al_2O_3	13.4	19.9	19.6	19.7
Fe_2O_3 (tot)	7.67	5.51	5.69	6.44
MnO	0.13	0.12	0.1	0.04
MgO	8.72	2.92	2.65	1.73
CaO	8.57	3.76	3.93	1.63
Na_2O	2.4	4.92	4.99	6.12
K_2O	1.4	4.29	4.21	4.1
P_2O_5	0.19	0.56	0.54	0.37
LOI	10.46	2.44	2.54	2.58
SO_3	0.05	0.63	0.37	0.15
Total	100.4	99.69	100.13	100.23
mg#	73	55	52	39
K_2O/Na_2O	0.58	0.87	0.84	0.67
V	200	170	170	160
Ni	166	4	4	4
Cu	112	5320	2850	750
Zn	103	92	86	175
Rb	45	75	75	100
Sr	790	1350	1250	840
Y	10	19	19	30
Zr	50	90	90	110
Nb	3	2	3	10
Ba	350	1200	1050	460
La	13.6	16.3	16.4	18.9
Ce	26	29	31	26
Yb	1.1	1.6	1.8	2.8
Hf	1.6	1.8	1.8	2.7

and plagioclase phenocrysts. This pervasive alteration style, which commonly obliterates primary volcanic textures, is developed only on a local scale, and is apparently related to major structures and/or contact zones of the intrusive monzonites.

Hydrothermal sericite associated with mineralization yields an $^{40}Ar/^{39}Ar$ step-heating age of 439.2 ± 1.2 Ma (Perkins et al. 1990a). U-Pb dating by ion microprobe on magmatic zircons from diorites near the Goonumbla deposits yields an age of 438 ± 3.5 Ma (Perkins et al. 1990b). These ages indicate that mineralization at the

119086 E26N Monzonite	119087 E26N Monzonite	119088 E27 Latite	119089 E26N Trachyte
64.6	64.9	60.1	60.2
0.36	0.36	0.48	0.52
17.5	17.1	18.4	16.4
1.89	2.13	4.6	3.49
0.03	0.03	0.18	0.05
1.38	1.04	1.45	1.11
0.87	1.15	2.16	1.57
5.76	6.39	5.93	4.6
4.68	3.99	5.31	6.82
0.18	0.15	0.28	0.24
2.01	1.83	1.48	3.71
0.79	1.29	0.02	2.22
100.12	100.35	100.35	100.96
63	53	42	43
0.81	0.62	0.89	1.48
120	110	80	110
5	4	4	4
3560	4150	33	9500
77	68	120	81
90	70	85	90
680	650	450	830
16	12	20	20
100	95	125	110
5	2	7	8
810	850	800	800
14.3	13	22.4	17.3
22	20	31	26
1.3	1.3	2.1	1.9
1.9	1.8	3.1	3.4

Goonumbla district was synchronous with magmatism and broadly contemporaneous with the formation of other gold deposits in the Lachlan Fold Belt (e.g. Glendale and Sheahan-Grants) as dated by Perkins et al. (1992).

Petrography and Geochemistry of the Shoshonitic Host Rocks. The Goonumbla igneous complex comprises mainly volcanic rocks with porphyritic textures which are intruded by monzonite stocks with equigranular textures.

Fig. 6.12.   Specimens from the Goonumbla igneous complex, New South Wales. (a) Quartz-stockwork veining in monzonite (119087). (b) Highly altered and mineralized monzonite (119092). (c) Microphotograph (plane polarized light) showing monzonite affected by propylitic alteration (119090) [FOV 5 mm]. Sample numbers refer to specimens held in the Museum of the Department of Geology and Geophysics, The University of Western Australia.

The grey andesitic host rocks consist mainly of labradorite as euhedral crystals up to 5 mm long and commonly intensely affected by saussuritization or dusting by sericite, with minor olivine and apatite microphenocrysts. The groundmass consists of plagioclase and orthoclase (Müller et al. 1994). The latites consist of phenocrysts of labradorite, orthoclase, clinopyroxene (augite), and apatite microphenocrysts which are set in a fine-grained groundmass of quartz, orthoclase, and plagioclase. The trachytes consist mainly of orthoclase, labradorite, augite, and mica phenocrysts set in a groundmass of orthoclase, plagioclase, and quartz. The trachytes commonly contain apatite microphenocrysts.

The intrusive, medium-grained, pink, quartz-rich monzonite bodies are holocrystalline rocks, comprising principally orthoclase, plagioclase, quartz, and mica. Secondary veins contain hydrothermal sericite.

Whole-rock analyses cover the petrographic spectrum from basaltic to dacitic rocks (i.e. SiO_2 contents between 46.8 and 64.9 wt %), as shown in Figure 6.13 and Table 6.6. The samples are characterized by high, but variable, Al_2O_3 contents (13.4–19.9 wt %), very high K_2O contents (up to 6.8 wt %) and high K_2O/Na_2O ratios

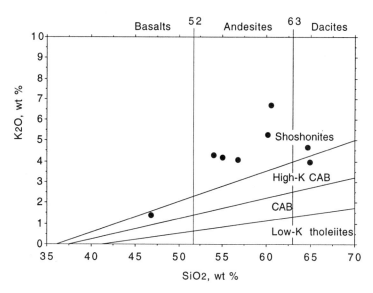

Fig. 6.13. SiO_2 versus K_2O plot (after Peccerillo and Taylor 1976a) for samples of the Goonumbla igneous complex, New South Wales. CAB = calc-alkaline basalt.

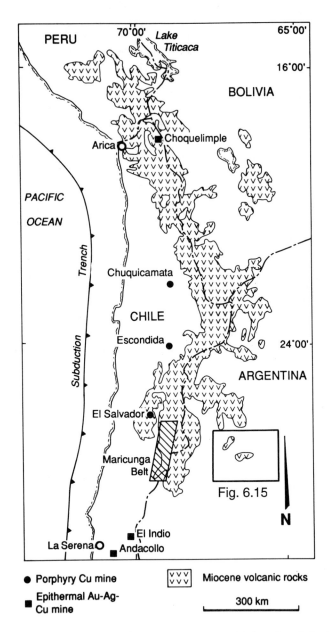

Fig. 6.14.   Geographic overview of the major copper-gold deposits of the Chilean and Argentinian Andes that are hosted by potassic igneous rocks. Modified after Vila and Sillitoe (1991).

(0.58–1.48) which are typical for the shoshonite association (Morrison 1980). The rocks also have enriched LILE concentrations (e.g. up to 1200 ppm Ba, 1350 ppm Sr), low HFSE contents (< 0.67 wt % TiO_2, < 125 ppm Zr, < 10 ppm Nb, < 3.4 ppm Hf) and very low LREE abundances (< 22.4 ppm La, < 31 ppm Ce).

Low mg# (< 63) suggest that the rocks were generated from evolved magmas, probably affected by clinopyroxene-biotite±apatite fractionation. However, one cross-cutting basaltic dyke has a higher mg# of 73.

All volcanic and intrusive rocks plot in the field of late oceanic-arc magmas in Figures 6.5 and 6.6.

6.4   Continental Arc Associations

Epithermal gold deposits hosted by high-K igneous rocks are known from both the North American Cordillera and the South American Andes (e.g. Clark 1993). The Andes are considered in more detail here because the geochemistry of the host rocks is better documented in the literature (see Sects 6.4.1 and 6.4.3). Generally, the host rocks of porphyry copper deposits in continental arcs of the western hemisphere are more felsic than those in late oceanic arcs of the Southwest Pacific (Titley 1975). This might be explained by a greater role of crustal assimilation during magma uprise in continental arcs.

Although there are exceptions, porphyry copper deposits in continental arcs commonly are more Mo-rich but Au-poor than those in late oceanic arcs (Sillitoe 1979).

6.4.1   Bajo de la Alumbrera Copper-Gold Deposit, Catamarca Province, Argentina

Introduction. The Bajo de la Alumbrera porphyry copper-gold deposit is located in the Catamarca Province in northwest Argentina (Figs 6.14, 6.15). The closest towns are Belen, about 100 km to the southwest, and Tucuman, about 300 km to the northeast.

In full production, Bajo de la Alumbrera will be the ninth largest copper mine in the world, and one of the biggest gold producers in South America. Total proven and probable reserves are 806 million tonnes at 0.53 wt % Cu and 0.64 ppm Au (D. Keough, pers. comm. 1998).

The deposit derives its name (English translation: below the edge) from its topographic location — it forms an oval-shaped topographic low below the edges of the unaltered, and thus more resistant, units of the surrounding Farallon Negro Formation.

Regional Geology. Northwest Argentina consists of three geological provinces: the Puna, Cordillera Oriental, and Sierras Pampeanas. The potassic igneous rocks that

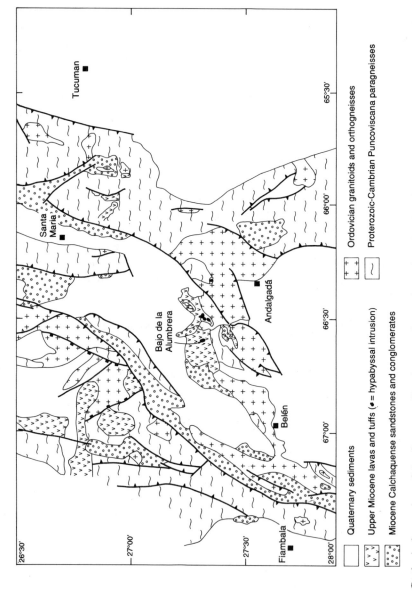

Fig. 6.15. Geological overview of the Catamarca Province, northwest Argentina, showing the location of the Bajo de la Alumbrera copper-gold deposit. Modified after Allmendinger et al. (1983), Allmendinger (1986), and Strecker et al. (1989).

host the Bajo de la Alumbrera deposit belong to the Farallon Negro Formation and are situated in the northernmost Sierras Pampeanas (Figs 6.14, 6.15).

The Sierras Pampeanas comprise a series of subparallel, north-trending, reverse fault-bounded, mountain ranges of Upper Proterozoic to Lower Paleozoic basement rocks which are separated by wide flat valleys in the foreland of the Central Andes (Caelles 1979). The thrusts and reverse faults commonly dip at about 60° to the east and have throws in access of 7000 m (Strecker et al. 1989). The basement rocks are predominantly amphibolite-facies mica schists and paragneisses which, in places, have been intruded by Upper Ordovician to Silurian granitoids (Caelles 1979). The Upper Ordovician to Silurian period in the Sierras Pampeanas was characterized by widespread plutonic activity and compressional deformation. This is consistent with modern plate-tectonic reconstructions of that time, implying collision between the eastern part of the North American continent and the western part of the South American continent during the Ordovician (Dalziel 1995).

The basement blocks have been uplifted along reverse faults during the Pliocene and early Pleistocene (Strecker et al. 1989), thus forming the intramontane basins of both the Puna and Sierras Pampeanas of northwest Argentina (Schwab and Lippolt 1974). The Pampean structural style is similar to that of the Laramide Ranges in western North America (Kay and Gordillo 1994), and has been suggested to be characteristic of foreland deformation over shallow subduction zones (Jordan and Allmendinger 1986; Introcaso et al. 1987).

In the Andes, the subduction angle varies along strike, and Recent volcanism is restricted to zones where the dip is about 30° (Kay et al. 1988). Miocene to Pliocene magmatism in the Sierras Pampeanas has been derived from the so-called "flat-slab zone", which represents the segment of the subducting oceanic Nazca Plate between 28° and 33° S (Kay et al. 1987; Kay and Abbruzzi 1996). In this zone, the subduction angle increasingly shallowed during the Miocene and Pliocene, beginning at ca. 18 Ma (Caelles 1979), probably caused by the subduction of a bend in the Juan-Fernandez Ridge (Pilger 1984). The shallowing subduction angle was accompanied by a progressive enrichment of the mantle wedge in LILE, and an increase in Pb and ΣNd values, resulting in enrichment of melts from the Upper Miocene relative to those from the Lower Miocene (Kay and Abbruzzi 1996). Quaternary and present-day volcanism is lacking in the flat-slab zone (Kay and Gordillo 1994).

The present subduction angle in this zone has been estimated to be about 10°, in contrast to the normal angle of about 30° to the north and south of the flat-slab zone (Caelles 1979) where Quaternary volcanism has been recorded (e.g. Strecker et al. 1989). The seismic transition between the flat-slab zone and the more steeply dipping subduction zone further to the north occurs between 25° and 28° S, and is characterized by a flexure below the 100-km depth contour for the Benioff Zone (Strecker et al. 1989). The Farallon Negro Formation, which hosts Bajo de la Alumbrera, is situated within this "transitional zone", immediately to the north of the shallowly dipping segment where the shallow subduction angle gradually increases again to 30° (Caelles 1979).

The emplacement of the Cenozoic potassic igneous complexes of northwest Argentina, such as the Farallon Negro Formation, was controlled by major northwest-

Table 6.7.   Major-element analyses (in wt %) of dacitic rocks from the Bajo de la Alumbrera porphyry copper-gold deposit, Argentina. From Müller and Forrestal (1998).

Province/deposit:	Bajo de la Alumbrera	Bajo de la Alumbrera	Bajo de la Alumbrera	Bajo de la Alumbrera
Location:	Argentina	Argentina	Argentina	Argentina
Rock type:	Dacite	Dacite	Dacite	Dacite
Tectonic setting:	Continental arc	Continental arc	Continental arc	Continental arc
Reference:	Müller and Forrestal (1998)	Müller and Forrestal (1998)	Müller and Forrestal (1998)	Müller and Forrestal (1998)
SiO_2	69.90	67.30	65.10	66.80
TiO_2	0.35	0.60	0.51	0.30
Al_2O_3	10.30	15.90	15.00	8.82
Fe_2O_3	2.45	4.40	5.81	4.49
FeO	3.36	0.69	1.19	7.12
MnO	0.15	0.01	0.01	0.07
MgO	0.99	0.77	0.83	0.92
CaO	1.56	0.09	0.30	0.91
Na_2O	0.80	0.20	0.20	0.45
K_2O	5.81	4.32	3.85	5.67
P_2O_5	0.10	0.11	0.04	0.01
LOI	3.98	5.49	6.88	4.29
Total	99.75	99.88	99.72	99.85
mg#	27	26	21	15
V	79	93	81	195
Cr	124	69	62	142
Ni	7	6	4	6
Rb	113	87	75	74
Sr	181	30	24	158
Y	8.6	6.0	6.0	5.6
Zr	78	111	115	72
Nb	11	3	3	7
Ba	504	243	134	487
La	9	32	31	10
Ce	18	61	65	18
Yb	0.8	2.1	1.2	0.7
Hf	2.9	3.9	4.0	2.4

striking lineaments (Allmendinger et al. 1983; Salfity 1985). Allmendinger et al. (1983) postulated a sinistral slip movement of up to 20 km along the lineaments, which are characterized by small negative gravity anomalies (Gotze et al. 1987), probably related to young volcanic activity which extends along these structures far to the east of the main volcanic chain (Schreiber and Schwab 1991).

Nature of Porphyry Copper-Gold Mineralization. Bajo de la Alumbrera is characterized by a classical alteration zonation, ranging from a potassic core which typically contains the highest gold grades (~ 2 ppm Au), surrounded by propylitization and an annular phyllic overprint (Müller and Forrestal 1998). The central zone with potassic alteration crops out on surface, and was the focal point when open-pit mining commenced in early 1997. The potassic alteration zone is characterized by very abundant magnetite (up to 10 vol. %) occurring both disseminated and as small veins, and by late-stage anhydrite veining, thus indicating the oxidized nature of the alteration and, presumably, the ore fluids derived from the melts.

The ore minerals are mainly pyrite and chalcopyrite with minor enargite and bornite. Metallurgical studies have shown that native gold occurs mainly as inclusions within pyrite and chalcopyrite. Interestingly, no gold occurs within the structure of the magnetite (S. Brown, pers. comm. 1994), which is a known host mineral of gold in many porphyry copper-gold deposits in the Philippines (I. Kavalieris, pers. comm. 1996).

Petrography and Geochemistry of the Potassic Host Rocks. The Upper Miocene (i.e. 6-10 Ma) Farallon Negro Formation forms an igneous complex of lava flows, breccia tuffs, and agglomerates, which, in places, are intruded by comagmatic hypabyssal stocks and domes (Müller and Forrestal 1998). Petrographically, both lava flows and tuffs vary from basaltic through latitic to rhyolitic in composition. The comagmatic intrusions (ca. 7.9 Ma) consist of monzonite stocks and dykes and hypabyssal trachyandesites and dacites (Stults 1985). The entire volcanic succession represents the basal remnants of a large, formerly up to 6 km high, stratovolcano some 16 km in diameter (Llambias 1972), and covers about 700 km^2 (Caelles 1979). The complex is situated within a tectonic depression bounded by uplifted Lower Paleozoic basement-rocks of the Sierra de Quilmes to the north and the Sierra de Aconquija to the east. The igneous rocks were erupted through a crystalline basement comprising amphibolite-facies metasedimentary rocks and granitoid batholiths (Müller and Forrestal 1998).

Porphyry copper-gold mineralization at Bajo de la Alumbrera was coeval with the emplacement of the hypabyssal dacite dome (H. Salgado, pers. comm. 1994), and several other prospects for porphyry copper-gold (e.g. Cerro Durazno, Agua Rica) and epithermal gold (e.g. Cerro Atajo) are known within the Farallon Negro Formation (Müller and Forrestal 1998). The hypabyssal dacite has a porphyritic texture consisting of phenocrysts of plagioclase, quartz, and biotite, with minor amphiboles and accessory apatite and magnetite, within a fine-grained groundmass mainly comprising feldspar and quartz. The accessory magnetite is indicative of the oxidized nature of the potassic melts (Müller and Forrestal 1998). The common presence of hydrous minerals, such as biotite and apatite, is consistent with the high volatile content of the magmas. Proximal to the orebody, the rocks are overprinted by intense secondary biotite-magnetite alteration, in which the magnetite content of the dacites may be as high as 10 vol. % (S. Brown, pers. comm. 1994).

The lava flows and pyroclastic deposits of the Farallon Negro Formation have typical high-K calc-alkaline compositions, whereas the shallow comagmatic dacite intrusion, which is genetically associated with the mineralization at Bajo de la Alumbrera, has a shoshonitic composition (Müller and Forrestal 1998). The dacites are characterized geochemically (Table 6.7) by rather low, but variable, Al_2O_3 contents (8.82–15.90 wt %). Fractional crystallization resulted in low mg# (< 27) and low concentrations of mantle-compatible trace-elements (e.g. < 142 ppm Cr, < 7 ppm Ni). The rocks are characterized by high concentrations of LILE (e.g. up to 5.81 wt % K_2O; up to 113 pm Rb, 181 ppm Sr, 504 ppm Ba), moderate LREE (e.g. up to 32 ppm La, 65 ppm Ce), and low HFSE (e.g. < 0.60 wt % TiO_2, < 115 ppm ppm, < 11 ppm Nb). They also have very high concentrations of LILE (e.g. up to 5.3 wt % K_2O at 62.4 wt % SiO_2), which are typical for potassic igneous rocks derived in a mature continental-arc setting (Figs 6.2–6.5, 6.16).

6.4.2 Bingham Copper Deposit, Utah, USA

Introduction. The Bingham porphyry copper deposit (Fig. 6.17) is located about 32 km southwest of Salt Lake City, in the central Oquirrh Mountains, Utah (Bowman et al. 1987; Waite et al. 1997). With a production of 1300 million tonnes of 0.85 wt % copper ore from 1904 to 1976, the Bingham mining district represents the largest porphyry copper deposit of North America and the world's largest skarn copper deposit with significant by-products of molybdenum, gold, and silver (Einaudi 1982).

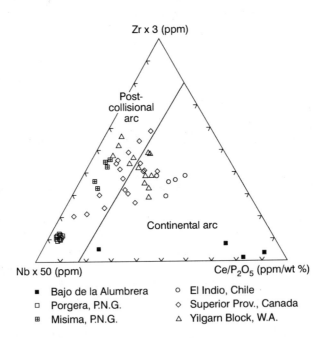

Fig. 6.16.   (Zrx3)-(Nbx50)-(Ce/P_2O_5) triangular diagram (see Chap. 3) showing only samples from continental and postcollisional arcs. Samples from the latter tectonic setting show some overlap due to P mobilization in altered Archaean shoshonitic lamprophyres. Adapted from Müller and Groves (1993).

Regional Geology. The igneous rocks of the Bingham complex (Fig. 6.17) intrude a sequence of limestones, quartzites, and related sedimentary rocks of the Upper Pennsylvanian Oquirrh Group (Bowman et al. 1987). The sedimentary basin in which the rocks accumulated is known as the Oquirrh Basin and contains up to 7.5 km of Late Pennsylvanian to Early Permian sedimentary rocks (Jordan and Douglas 1980). Folding of the sedimentary rocks is characterized by large, open, northwest-striking anticlines and synclines (Lanier et al. 1978a). The dominant fold within the mining area strikes northwest and plunges 45° northwest, with the igneous intrusions and

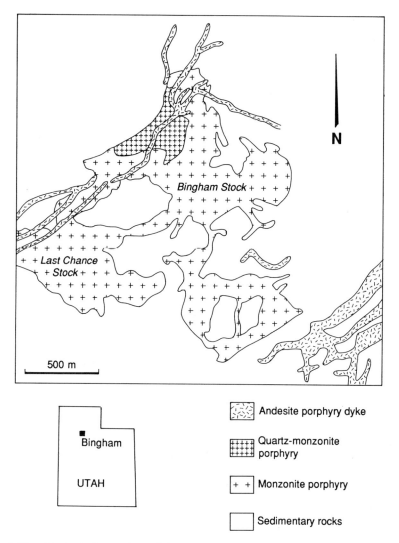

Fig. 6.17.   Geological overview of the Bingham porphyry copper deposit, Utah, USA. Modified after Warnaars et al. (1978).

Table 6.8. Major-element analyses (in wt %) of potassic igneous rocks from the Bingham porphyry copper deposit, Utah. Fe_2O_3 (tot) = total iron calculated as ferric oxide. From Waite et al. (1997).

Province/deposit:	Bingham	Bingham	Bingham
Location:	Utah, USA	Utah, USA	Utah, USA
Rock type:	Latite	Dacite	Quartz monzonite
Tectonic setting:	Continental arc	Continental arc	Continental arc
Reference:	Waite et al. (1997)	Waite et al. (1997)	Waite et al. (1997)
SiO_2	61.54	62.89	57.88
TiO_2	0.69	0.69	0.99
Al_2O_3	14.81	14.50	15.20
Fe_2O_3 (tot)	6.19	6.10	7.29
MnO	0.09	0.09	0.11
MgO	4.23	4.21	4.38
CaO	5.31	4.49	6.31
Na_2O	3.14	2.98	3.38
K_2O	3.67	3.76	3.96
P_2O_5	0.34	0.32	0.49
LOI	1.30	2.37	0.35
Total	100.01	100.02	99.99
mg#	61	61	58
V	142	130	161
Cr	166	853	28
Ni	52	45	31
Rb	105	216	105
Sr	708	203	1199
Y	20	15	20
Zr	229	228	219
Nb	13	16	9
Ba	2074	2100	2518
La	76	99	92
Ce	117	152	159

porphyry copper mineralization localized along the southwestern limb (Bowman et al. 1987).

Nature of Porphyry Copper Mineralization. Porphyry copper±gold mineralization probably took place during several pulses of hydrothermal activity which started at 39.8 Ma and probably culminated prior to emplacement of late-stage latite dykes (see below) at 38.0 Ma (Warnaars et al. 1978). Mineralization is focussed on a hydrothermally altered quartz-monzonite stock, first referred to as the Bingham stock by Butler (1920). Three alteration zones can be distinguished (Lanier et al. 1978a):

- Internal potassic alteration zone.
- Surrounding propylitic alteration zone.

– Late-stage and mainly fault-controlled sericitic alteration zone.

Porphyry-type mineralization occurs mainly within the potassic alteration zone which consists of quartz, potassic feldspar, hydrothermal biotite, and sulphide minerals (Bowman et al. 1987). Mineralization consists of a high-grade core with pyrite and chalcopyrite, a molybdenite stockwork zone, and an outermost pyrite zone (Bowman et al. 1987). The surrounding propylitic alteration zone partially overlaps potassic alteration and its typical minerals include actinolite, chlorite, and epidote (Bowman et al. 1987; Keith et al. 1997). The late-stage sericitic alteration comprises sericite replacing plagioclase and mafic minerals, and is developed most prominently in the western part of the Bingham deposit (Bowman et al. 1987). Fluid inclusion studies indicate both magmatic and meteoric components in the hydrothermal fluid (Bowman et al. 1987).

Petrography and Geochemistry of the Potassic Host Rocks. Small amounts of minette, lamproite, and shoshonite magmas accompanied the calc-alkaline Eocene-Oligocene magmatism in central Utah. Studies by Keith et al. (1997) suggest that the latites and monzonites at Bingham are the products of magma mixing of shoshonites and minettes with intermediate melts.

The ages of the intrusive rocks from the Bingham stock range from 39.8 ± 0.4 to 38.8 ± 0.4 Ma (Warnaars et al. 1978). Two phases of the monzonite intrusion can be distinguished (see Fig. 6.17) based on their normative quartz contents: an older (39.8 ± 0.4 Ma), dark grey, monzonite phase with less quartz (< 10 vol. %); and a younger (38.8 ± 0.4 Ma), light grey, quartz-monzonite phase with abundant quartz (> 20 vol. %). The older monzonite is fine grained and equigranular, and consists of potassic feldspar, plagioclase, quartz, augite, phlogopite, and minor amphibole (Lanier et al. 1978a): it forms the main host for porphyry copper mineralization (Lanier et al. 1978b). The younger quartz-monzonite forms a northeast-tending dyke up to 400 m thick, and represents the major part of the Bingham stock. In places, the monzonite stocks are cut by hypabyssal andesite porphyry dykes (Fig. 6.17), which belong to a swarm of northeast-trending dykes and sills intruding the northwest margin of the Bingham stock (Lanier et al. 1978a). These latite dykes are dated at ca. 38.0 ± 0.2 Ma (Warnaars et al. 1978), and comprise large phenocrysts of plagioclase, potassic feldspar (both up to 5 mm long), biotite, and quartz, and apatite microphenocrysts (Wilson 1978). A high halogen (i.e. Cl and F) content of the magma is suggested by fluorapatite in the groundmass and the high salinity of fluid inclusions (Wilson 1978).

Table 6.8 shows representative whole-rock analyses for the potassic igneous rocks from Bingham (Waite et al. 1997). The samples are characterized by very high LILE concentrations (e.g. up to 3.96 wt % K_2O; up to 216 ppm Rb, 1199 ppm Sr, 2518 ppm Ba), moderate LREE concentrations (e.g. up to 99 ppm La, 159 ppm Ce), and low HFSE abundances (e.g. < 0.99 wt % TiO_2, < 229 ppm Zr, < 16 ppm Nb). Fractional crystallization resulted in relatively low mg# (< 61), and relatively low concentrations of mantle-compatible trace-elements (e.g. < 161 ppm V, < 52 ppm Ni). The rocks have elevated P_2O_5 concentrations (up to 0.49 wt %), which is consistent with the presence of apatite microphenocrysts (Wilson 1978).

The samples have not been plotted on discrimination diagrams due to the lack of Hf analyses. However, based on their geochemical compositions, the rocks are interpreted to be derived in a mature continental setting (Müller et al. 1992b).

6.4.3 El Indio Gold Deposit, Chile

Introduction. The Chilean Andes represent the largest copper province in the world (Sillitoe and Camus 1991). Since the discovery of the epithermal gold deposits El Indio, Tambo, Choquelimpie, El Hueso, and La Coipa (Fig. 6.18), Chile has also earned the status of a major gold province (Sillitoe 1991; Clark 1993). Most of the 18 principal epithermal- or porphyry-type gold deposits with more than 10 tonnes of contained gold are associated with Miocene stratovolcanoes and/or dome complexes (Sillitoe 1991). These gold deposits are concentrated in northern and central Chile, commonly at altitudes in access of 4000 m above main sea level (MSL) (Sillitoe 1991).

El Indio is one of a number of deposits related to high-K rocks. The El Indio deposit, together with its neighbouring deposits, Tambo and Nevada, is one of the most important gold-silver producers in Chile. El Indio is known for its high-grade mineralization, with about 2 million ounces of gold, 6 million ounces of silver, and 100000 tonnes of copper produced since operations began in 1980 (Jannas et al. 1990). Reported reserves (June, 1989) contain 2.5 million ounces of gold, 18.8 million ounces of silver, and 306700 tonnes of copper (Jannas et al. 1990).

El Indio (Fig. 6.19) is located about 500 km north of Santiago and 180 km east of La Serena in a rugged, mountainous terrain close to the Argentinian border. The mine workings reach 4150 m above sea level (Siddeley and Araneda 1986; Jannas et al. 1990).

Regional Geology. The Chilean copper-gold deposits are part of a series of linear volcano-plutonic arcs that were generated progressively farther eastward during subduction of the oceanic Nazca plate beneath the continental South American plate from the Mesozoic to Recent (Sillitoe 1991). Most of the copper-gold deposits hosted by calc-alkaline and shoshonitic rocks are Upper Cretaceous to Miocene in age (Davidson and Mpodozis 1991; Gropper et al. 1991; Reyes 1991; Clark 1993). The Cretaceous-Upper Tertiary volcanic and pyroclastic rocks in the high Andes of north-central Chile form a regional north-south-trending belt over 150 km long which is underlain by Carboniferous to Lower Triassic calc-alkaline granodioritic batholiths (Fig. 6.18; Jannas et al. 1990). These basement rocks have been thrust over the Cretaceous-Tertiary volcanic rocks by the steeply west-dipping Baños del Toro Fault, which defines the western boundary of the volcanic belt (Jannas et al. 1990).

El Indio is located within the Upper Oligocene to Lower Miocene Doña Ana Formation (Fig. 6.18), which belongs to the volcanic belt of the High Cordillera.

Mineralization and argillic alteration at El Indio are genetically related to hypabyssal dioritic and monzodioritic stocks and dykes (Kay et al. 1987; Jannas et al. 1990). These shallow intrusions belong to the Infiernillo Unit (ca. 16.7 Ma), which intruded the bimodal andesite-rhyolite sequences (ca. 18-27 Ma) of the Doña Ana Formation (Kay et al. 1987, 1988). However, the main host rocks of the mineralization at El Indio are pyroclastic units with dacitic compositions (Jannas et al. 1990).

The regional structural setting of El Indio is dominated by north-south-trending faults parallel to the main Andean trend (Walthier et al. 1985).

The major part of the recognized gold-silver±copper mineralization occurs within a structural block only 150 m wide by 500 m long. This block contains more than 40 mineralized veins and is bounded by two northeast-striking principal faults, which

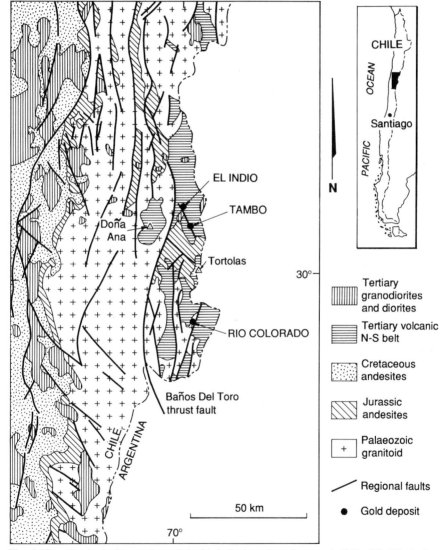

Fig. 6.18. Geological overview of the high Andes of north-central Chile. Modified after Siddeley and Araneda (1986).

dip steeply to the northwest (Jannas et al. 1990). Individual veins can extend more than 400 m and their thicknesses vary between a few centimetres and 20 metres (Jannas et al. 1990). The veins are normally lenticular (Siddeley and Araneda 1986).

Nature of Epithermal Gold Mineralization. The epithermal gold deposits in Chile (Sillitoe 1991) include both high- and low-sulphidation types, and range from vein-dominated systems at El Indio through hydrothermal breccia-dominated systems at

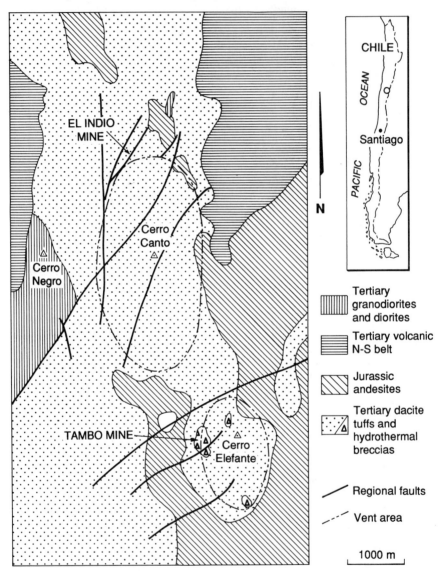

Fig. 6.19. Geological overview of the El Indio and Tambo gold deposits, Chile. Modified after Siddeley and Araneda (1986).

Choquelimpie, to complex systems combining stockworks, disseminated zones, breccias, and veins, which are comparable to the Goonumbla porphyry copper-gold deposit (see Sect. 6.3.4).

El Indio is characterized by argillic alteration and silicification (Jannas et al. 1990), both of which are typical features for many high-sulphidation epithermal gold deposits (e.g. White et al. 1995). The argillic alteration affects mainly the dacitic pyroclastic units, converting feldspars to kaolinite, sericite and dickite, with minor pyrophyllite and montmorillonite (Siddeley and Araneda 1986). The silicification occurs in patchy, pervasive zones, locally obliterating wallrock textures, and it predominates in the vicinity of high-grade ores (Jannas et al. 1990). The replacement of feldspars by alunite and jarosite is a minor, late-stage feature (Siddeley and Araneda 1986). Propylitic alteration is virtually absent at El Indio, although it is dominant outside the two major faults that bound the deposit (Walthier et al. 1985).

All mineralization is structurally controlled, along extensive tensional features, and two major ore types can be distinguished (Siddeley and Araneda 1986):

—   Massive sulphide veins.
—   Quartz-gold veins.

The bulk ore at El Indio is derived from the massive sulphide veins, which consist mainly of enargite and pyrite, together forming about 90 vol. % of the vein material (Siddeley and Araneda 1986). The remaining 10 vol. % of the massive sulphide veins consists of altered wallrock fragments.

The richer gold ores consist of younger quartz veins which, in places, cut the massive sulphide veins (Siddeley and Araneda 1986). Gold in these veins is predominantly in the native state, as thin trails or disseminated within the quartz, and can be accompanied by minor pyrite, enargite, and tennantite (Siddeley and Araneda 1986). The average gold content of these quartz veins is between 18 and 25 ppm, but locally exceeds 200 ppm (Jannas et al. 1990). The quartz gangue material is grey, cryptocrystalline, banded, colloform, cherty or drusy (Siddeley and Araneda 1986).

Petrography and Geochemistry of the Potassic Host Rocks. Small amounts of minette, lamproite, and shoshonite magmas accompanied the calc-alkaline Eocene-Oligocene magmatism in central Utah. Studies by Keith et al. (1997) suggest that the latites and monzonites at Bingham are the products of magma mixing of shoshonites and minettes with intermediate melts.

The dioritic source rocks to gold mineralization at El Indio belong to the Infiernillo Unit and have been dated at ca. 16.7 Ma (Kay et al. 1988, and references therein), with mineralization thought to be contemporaneous at ca. 16.0 Ma (Tschischow 1989).

The monzogranites and granodiorites consist of clino- and orthopyroxene, quartz, plagioclase, alkali feldspar, biotite and amphibole (Kay et al. 1987). Geochemically (Table 6.9), the rocks are characterized by high LILE concentrations (e.g. up to 4.0 wt % K_2O; up to 16 ppm Cs, 624 ppm Ba, 270 ppm Rb) and relatively low LREE contents (e.g. < 32 ppm La, < 73 ppm Ce) (Kay et al. 1987). Only limited data for the HFSE of the Infiernillo diorites have been published. However, whole-rock data

Table 6.9. Major- and trace-element analyses of monzodiorites from El Indio gold deposit, Chile. Major elements are in wt %, and trace elements are in ppm. Fe_2O_3 (tot) = total iron calculated as ferric oxide. From Tschischow (1989).

Province/deposit: Location: Rock type: Tectonic setting: Reference:	El Indio Chile Monzodiorite Continental arc Tschischow (1989)	El Indio Chile Monzodiorite Continental arc Tschischow (1989)	El Indio Chile Monzodiorite Continental arc Tschischow (1989)	El Indio Chile Monzodiorite Continental arc Tschischow (1989)	El Indio Chile Monzodiorite Continental arc Tschischow (1989)
SiO_2	68.53	63.24	65.29	61.51	61.76
TiO_2	0.75	0.89	0.62	0.76	0.77
Al_2O_3	16.69	16.53	16.39	17.99	17.11
Fe_2O_3 (tot)	4.86	5.03	3.97	5.07	5.27
MnO	0.08	0.07	0.06	0.08	0.09
MgO	2.92	2.16	1.90	2.32	2.65
CaO	3.05	4.57	4.13	5.68	5.31
Na_2O	4.66	4.07	3.83	3.44	3.82
K_2O	3.33	3.25	3.66	3.03	3.03
P_2O_5	0.13	0.17	0.14	0.12	0.18
LOI	n.a.	n.a.	n.a.	n.a.	n.a.
Total	100.00	99.88	99.99	100.00	99.99
mg#	58	50	53	52	54
V	n.a.	n.a.	n.a.	n.a.	n.a.
Cr	17	23	13	20	18
Ni	n.a.	n.a.	n.a.	n.a.	n.a.
Rb	151	166	179	150	168
Sr	n.a.	n.a.	n.a.	n.a.	n.a.
Y	n.a.	n.a.	n.a.	n.a.	n.a.
Zr	160	140	150	130	100
Nb	15	10	11	9	11
Ba	624	523	512	516	580
La	26	30	29	26	27
Ce	58	71	64	64	60
Th	n.a.	n.a.	n.a.	n.a.	n.a.
Ta	n.a.	n.a.	n.a.	n.a.	n.a.
Hf	4	7	5	6	6

from Tschischow (1989) reveal low Zr abundances (~ 150 ppm), slightly elevated, but variable, Nb (5–14 ppm), and intermediate Hf (~ 6 ppm). Their very high contents of Th (up to 36 ppm) and U (up to 7 ppm) suggest some crustal contamination during magma uprise (Kay et al. 1987).

El Indio is only one example of a Chilean gold deposit associated with potassic igneous rocks. In fact, many of the Chilean gold deposits, such as Andacollo (Reyes 1991), Choquelimpie (Gropper et al. 1991), and those of the Maricunga Belt (Dostal et al. 1977b; Vila and Sillitoe 1991), are hosted by potassic calc-alkaline or shoshonitic rocks interpreted to have formed during subduction in a continental arc setting (Levi et al. 1988; Clark 1993). In places, copper-gold mineralization is associated with widespread potassic alteration (J. Piekenbrock, pers. comm. 1994). However, whole-rock geochemical analyses from unaltered volcanic host rocks (e.g. Gropper et al. 1991) suggest that the original intrusions were characterized by high-K affinities *before* potassic alteration took place.

6.4.4 Twin Buttes Copper Deposit, Arizona, USA

Introduction. The Twin Buttes porphyry copper deposit is located about 40 km south of Tucson in the foothills of the eastern Sierrita Mountains, in the Pima mining district, Arizona, and ranks as one of the major copper producers of the USA (Barter and Kelly 1982). Open-pit mining at Twin Buttes commenced in mid-1965, and sulphide exploitation began in 1969 (Barter and Kelly 1982).

Regional Geology. The Pima mining district is mainly composed of Paleozoic metasedimentary and acid to intermediate igneous rocks of Triassic through Miocene, as well as Precambrian, age (Cooper 1973). The Paleozoic fluvial clastic rocks were originally deposited in a shallow shelf and marine environment, and later folded and metamorphosed during the lower Cretaceous; they commonly strike northwest and have vertical to moderate northeasterly dips (Barter and Kelly 1982). Northeasterly fracturing is a significant mineralization control in the Pima district and apparently influences at least the later stages of mineralization at Twin Buttes (Barter and Kelly 1982). A set of southeast-trending intrusions cut the Paleozoic metasedimentary rocks and occupy the central portion of the Twin Buttes mineralized zone.

Nature of Porphyry Copper Mineralization. Porphyry copper mineralization at Twin Buttes is believed to be genetically related to a contemporaneous intrusive complex of quartz monzonites with minor granites and rhyodacites, and was formed during the Laramide Orogeny, in the Paleocene, along a major northwest-trending fault zone (Barter and Kelly 1982).

The intrusive complex is cut by three northeast-trending, near-vertical faults which segment the mineralized zone into five ore bodies: northwest, north porphyry contact, northeast, main, and arkose ore bodies. Porphyry copper-type mineralization consists mainly of pyrite and chalcopyrite with minor sphalerite, molybdenite, bornite, pyrrhotite, chalcocite, and galena, which occur as disseminations, in hairline fractures, or as stockwork-veining (Barter and Kelly 1982). Although there are economically significant amounts of silver in the mineralized zone, only small amounts of two silver minerals — tetrahedrite and matildite — have been identified (Salek 1976). The higher-grade hypogene mineralization (> 4000 ppm Cu) normally occurs in the rhyodacite and the metasedimentary rocks, whereas the quartz monzonites have lower grades (< 2000 ppm Cu; cf. Barter and Kelly 1982). The highest ore grades are in zones of pervasive potassic alteration, with the development of sericite and potassic feldspar as halos along quartz-sulphide veins (Barter and Kelly 1982).

Petrography of the Potassic Host Rocks. Two types of the Paleocene monzonitic host rocks of Twin Buttes can be distinguished (Barter and Kelly 1982):

– A younger porphyritic quartz-monzonite which contains very large phenocrysts of orthoclase (up to 4 cm long), biotite (up to 2 mm long) and titanite (up to 1 mm long) in a medium-grained groundmass of quartz, plagioclase, microcline, and orthoclase. Biotite phenocrysts from the porphyritic quartz-monzonite have been dated at 57.1 ± 2 Ma (Kelly 1977).
– An older, finer-grained quartz-monzonite variety which can be further subdivided into three types:
  1. An older aphanitic sub-type which has been dated at 58.6 ± 2.5 Ma (Kelly 1977), and comprises phenocrysts of plagioclase, orthoclase, quartz, and biotite in an extremely fine-grained (i.e. aphanitic) groundmass of quartz and orthoclase.
  2. A xenolithic sub-type which contains abundant xenoliths of Paleozoic metasedimentary rocks.
  3. A younger aplitic sub-type which consists of phenocrysts of quartz, plagioclase, and orthoclase in a fine-grained groundmass of quartz and orthoclase (Barter and Kelly 1982).

Other types of intrusive rocks observed in the mineralized zone at Twin Buttes include dacite, hornblende-quartz monzonite porphyry, and a Miocene andesite dyke (Barter and Kelly 1982).

Unfortunately, no representative geochemical whole-rock data for the igneous host rocks from Twin Buttes were available to the authors, but the petrography of the rocks (i.e. orthoclase and biotite phenocrysts) indicates that they are potassic.

6.5   Postcollisional Arc Associations

6.5.1   Grasberg Copper-Gold Deposit, Indonesia

Introduction. The Pliocene Grasberg porphyry copper-gold deposit (Fig. 6.20) is situated in the Ertsberg Mineral District of Irian Jaya, at 4° S latitude in the western

(Indonesian) part of the New Guinea main island. The Ertsberg District is about 70 km inland from the Arafura Sea at an altitude of 3500–4500 m MSL (Van Nort et al. 1991). The area is one of only three in the world having permanent mountain glaciers at equatorial latitudes (MacDonald and Arnold 1994). The mineral potential of the area was first recognized in 1936, but systematic exploration by Freeport Mining Co. only commenced after 1960. The history of early exploration at Grasberg has been described in detail by Wilson (1981).

Estimates of the mineable reserve for the Grasberg deposit are 362 million tonnes at 1.53 wt % Cu, 1.97 ppm Au and 3.24 ppm Ag (Hickson 1991).

Regional Geology. The island of New Guinea has long been recognized as the product of Miocene collision between the north-moving Australian plate and the southwest-migrating Pacific plate (e.g. Puntodewo et al. 1994). Undeformed continental crust (Van Nort et al. 1991) of the Australian plate extends northward from beneath the Arafura Sea along the southern coastal plain to the southern edge of the Papuan Fold Belt, commonly referred to as the "highland mountainbelt" in the literature (e.g. Richards et al. 1991).

The Grasberg copper-gold mineralization is hosted by igneous intrusive rocks of intermediate composition (Fig. 6.20), ranging from diorite to quartz-monzonites (Van Nort et al. 1991), which intrude Tertiary limestones of the New Guinea Group (MacDonald and Arnold 1994). The limestones have been compressed into a series of tight isoclinal folds with nearly vertical axial planes. Pluton emplacement, as well as subsequent copper-gold mineralization, may have been controlled by the inter-

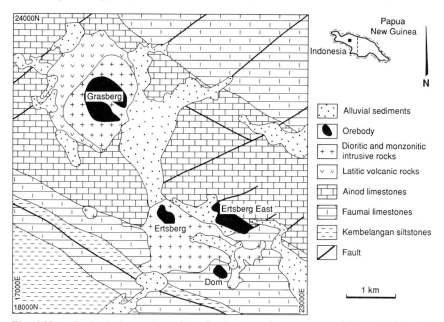

Fig. 6.20. Geological overview of the Grasberg porphyry copper-gold deposit, Indonesia. Modified after MacDonald and Arnold (1994).

sections of steep, northwest-trending reverse faults and northeast-trending sinistral strike-slip faults (MacDonald and Arnold 1994). The Grasberg intrusion is dated at ca. 3 ± 0.5 Ma using conventional K-Ar methods (MacDonald and Arnold 1994).

Nature of Porphyry Copper-Gold Mineralization. The primary hypogene sulphide mineralogy consists mainly of chalcopyrite, bornite, and minor pyrite (Van Nort et al. 1991). Chalcopyrite occurs mainly as fracture fillings throughout the porphyry, and as veinlets associated with quartz in the stockwork zone. The igneous rocks also contain disseminated magnetite (2–5 vol. %; MacDonald and Arnold 1994). Gold is closely associated with chalcopyrite and bornite, and gold values appear to increase with depth (Van Nort et al. 1991). The native gold is characterized by high Pd contents (up to 3500 ppm; I. Kavalieris, pers. comm. 1996). Supergene effects are minimal, with the development of only a weak, thin (about 5 m), leached capping characterized by secondary chalcocite, digenite, and covellite. However, the area has gone through several glacial ice advances over the past one million years, and most of the leached capping was probably scoured off (Van Nort et al. 1991).

A central potassic alteration zone, with associated quartz-stockwork, grades outward into phyllic alteration and then into a thin propylitic zone near the intrusive contact with the limestone country rocks (Hickson 1991). The potassic alteration zone is characterized by secondary, very fine-grained biotite and orthoclase (Van Nort et al. 1991). Anhydrite development is also characteristic and accompanies potassic alteration (MacDonald and Arnold 1994). Phyllic alteration surrounds the potassic zone, and consists mainly of sericite, kaolinite, and pyrite. Primary disseminated magnetite survives potassic alteration, but is not observed in the phyllic alteration zone (MacDonald and Arnold 1994). Propylitic alteration is not well developed at Grasberg and, where observed, rock textures are preserved but feldspar and mafic phenocrysts are altered to either clay or chlorite (Van Nort et al. 1991).

The bulk of the high-grade copper-gold mineralization is hosted by quartz-stockwork veins within the potassic alteration zone. However, major ore reserves are also represented by skarn-type mineralization formed at contact zones between igneous intrusions and the dolomitic country rocks (Hickson 1991).

Petrology and Geochemistry of the Potassic Host Rocks. The porphyry copper-gold mineralization in the Ertsberg Mineral District is hosted by two major and several minor intrusions of dioritic to monzonitic composition (MacDonald and Arnold 1994). The economically most important intrusions are the Grasberg, Ertsberg, and Dom stocks (Fig. 6.20). The Grasberg stock consists mainly of a medium-grained, porphyritic diorite comprising phenocrysts of plagioclase (55 vol. %), hornblende and augite (10 vol. %), and brown biotite (3-5 vol. %), with local quartz grains in a fine-grained groundmass of biotite, orthoclase, and plagioclase (Hickson 1991; Van Nort et al. 1991). Typical skarn ore bodies at Grasberg are chalcopyrite-bornite-magnetite-silicate-rich replacements of dolomitic country rocks (Hickson 1991).

Geochemically, the igneous host rocks at Grasberg are strongly evolved (Table 6.10), as reflected by high SiO_2 (> 59.5 wt %) and low MgO (< 3.4 wt %) contents, low mg# (< 60), and low concentrations of the mantle-compatible trace-elements

Table 6.10. Major- and trace-element analyses of potassic igneous rocks from the Grasberg porphyry copper-gold deposit, Indonesia (Freeport Mining Co., unpubl. data 1993). Major elements are in wt %, and trace elements are in ppm. Fe_2O_3 (tot) = total iron calculated as ferric oxide.

Province/deposit:	Grasberg	Grasberg
Location:	Indonesia	Indonesia
Rock type:	Monzonite	Diorite
Tectonic setting:	Postcollisional arc	Postcollisional arc
Reference:	Freeport Mining Co. (unpubl. data, 1993)	Freeport Mining Co. (unpubl. data, 1993)
SiO_2	59.45	66.75
TiO_2	0.38	0.50
Al_2O_3	10.79	13.74
Fe_2O_3 (tot)	15.57	3.22
MnO	0.05	0.01
MgO	3.35	2.09
CaO	0.42	0.62
Na_2O	1.16	1.20
K_2O	6.31	8.23
P_2O_5	0.27	0.23
LOI	1.32	2.57
Total	99.07	99.16
mg#	33	60
V	97	87
Cr	13	56
Ni	11	35
Rb	150	170
Sr	179	303
Y	13	11
Zr	86	168
Nb	9	15
Ba	277	1554
La	9	49
Ce	15	79
Hf	n.a.	n.a.

(e.g. < 100 ppm V, < 60 ppm Cr, < 35 ppm Ni). The rocks are generally characterized by very high K_2O contents (up to 8.4 wt %), high K_2O/Na_2O ratios (1.3–8.3), and high LILE (up to 248 ppm Rb, up to 1554 ppm Ba), intermediate LREE (< 50 ppm La, < 80 ppm Ce), and low HFSE (< 0.54 wt % TiO_2, < 180 ppm Zr) contents. The rocks contain relatively high Nb concentrations (up to 35 ppm), which are consistent with those of potassic igneous rocks from many other postcollisional arc settings. Overall, the potassic igneous rocks from Grasberg are similar to those from the Porgera gold deposit, Papua New Guinea (see Sect. 6.5.3). However, the Grasberg igneous rocks are more evolved, having lower Na_2O, CaO and TiO_2 contents, whereas those from Porgera are characterized by more primitive compositions.

Due to the limited data set (e.g. no Hf data) available to the authors, the rocks could not be plotted on the discrimination diagrams of Chapter 3. However, based on the whole-rock geochemistry of the Grasberg igneous suite, which is comparable to that of the potassic igneous rocks which host the gold mineralization at Porgera, it is interpreted to be the product of a postcollisional arc setting.

6.5.2 Misima Gold Deposit, Misima Island, Papua New Guinea

Introduction. The Misima gold deposit is located on Misima Island in the Louisiade Archipelago, some 670 km east-southeast of Port Moresby and 240 km east-southeast of the Papua New Guinea mainland (Lewis and Wilson 1990). Misima Island is about 40 km long and up to 9 km wide, and has a rugged topography (Wilson and Barwick 1991). The mine is located in the eastern part of the island (Fig. 6.21). Open-cut mining commenced in 1988, with mineable reserves of 55.9 million tonnes at 1.38 ppm Au and 21.0 ppm Ag at a cutoff grade of 0.7 ppm Au equivalent (Lewis and Wilson 1990).

Regional Geology. The geology of Misima Island (Fig. 6.21) is dominated by Eocene or older metamorphic rocks (Appleby et al. 1995), which have been subdivided into the higher-grade Awaibi Association and the lower-grade Sisa Association (Wilson and Barwick 1991). These two metamorphic units are separated by an extensional, shallow-angle detachment fault with a mylonitic fabric (Fig. 6.22; Appleby et al. 1995). The Awaibi Association amphibolite-facies metamorphic rocks structurally underlie the less-metamorphosed Sisa Association rocks of upper greenschist facies, which occur on the eastern part of the island and contain the gold mineralization (Wilson and Barwick 1991). The Sisa Association is a conformable sequence of dominantly psammopelitic schists enclosing fine-grained, massive to foliated greenstones (de Keyser 1961; Clarke et al. 1990).

The Sisa association units are gently dipping and are folded into an east-trending, east-plunging antiform (Wilson and Barwick 1991). The rocks are overprinted by a penetrative deformation which resulted in foliation, crenulation fabrics, and isoclinal to chevron folding (Clarke 1988). The schists and greenstones of the Sisa Association have been intruded by hypabyssal andesitic to dacitic stocks, sills, and dykes known as the Boiou microgranodiorite (Fig. 6.22; Clarke et al. 1990), which postdates the deformation (Clarke 1988). SHRIMP II U-Pb dating of magmatic zircons from this granodiorite yields a crystallization age of 8.1 ± 0.4 Ma (Appleby et al. 1995).

At least three phases of faulting can be identified, the best-known relationships being in the Umuna Fault Zone, a 50 to 200 m-wide zone of subparallel and anastomosing fractures (Lewis and Wilson 1990). The Misima gold deposit is situated within the Umuna Fault Zone (A.-K. Appleby, written comm. 1996). The epithermal gold mineralization at Misima is spatially associated with contemporaneous

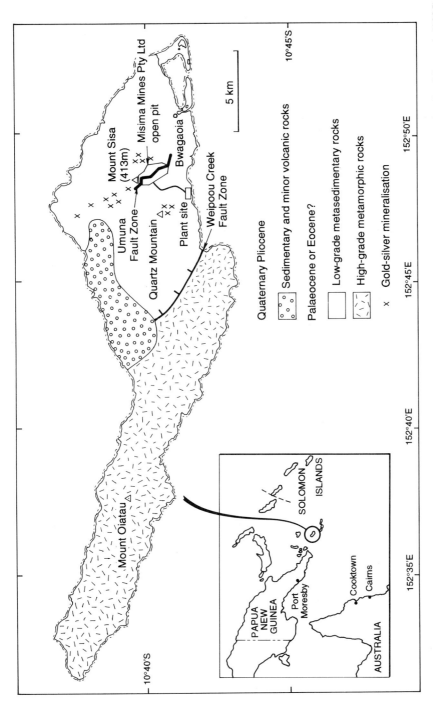

Fig. 6.21.   Geological overview of Misima Island, Papua New Guinea, showing gold-silver mineralization. Modified after Appleby et al. (1995).

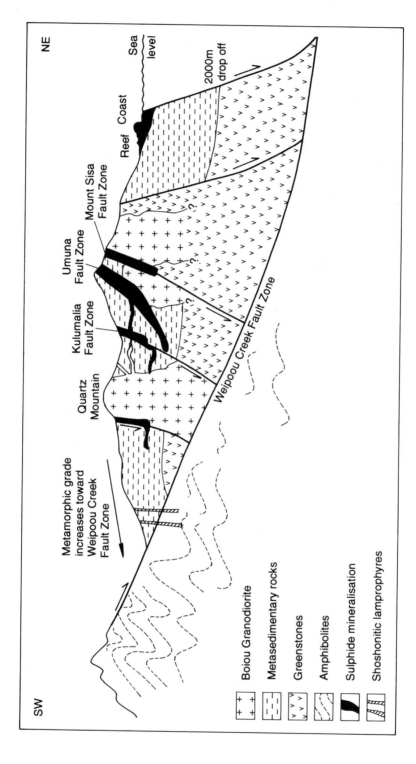

Fig. 6.22.   Geological cross-section of Misima Island, Papua New Guinea. Modified after Appleby et al. (1995).

shoshonitic lamprophyres, and both the mineralization and lamprophyre dykes have been K-Ar dated at 3.5 Ma (A.-K. Appleby, written comm. 1996).

Nature of Epithermal Gold Mineralization. Two mineralization events can be distinguished at Misima (Wilson and Barwick 1991):

— Early uneconomic *porphyry-style* pyrite-chalcopyrite-bornite-molybdenite mineralization with both weak biotite-magnetite and propylitic alteration associated with the Boiou microgranodiorite.
— Economic *epithermal* precious-metal mineralization with pyrite-sphalerite-galena-chalcopyrite±gold±silver associated with quartz and quartz-carbonate gangue as fracture fill. The epithermal mineralization is focused in high- and low-angle structures in the upper plate of the detachment fault (Appleby et al. 1995).

The north-northwest-striking Umuna Fault Zone is the most significant known locus of mineralization on Misima Island (Fig. 6.22). The > 5 million ounce gold orebody is about 100 to 200 m wide, 500 m deep, and has a 3 km strike length (Appleby et al. 1995). The high-grade gold mineralization is sited in en echelon dilational sites, indicating normal movement with dextral slip along the Umuna Fault Zone at the time of mineralization (Appleby et al. 1995).

The vein mineralogy is dominated by quartz or carbonate gangue with galena, sphalerite, pyrite, and chalcopyrite. Gold shows a distinctive correlation with galena (Appleby et al. 1995). The upper levels of the deposit are dominated by < 20 m-wide silica veins with minor barite, whereas carbonate veins and breccia infill are more common at depth (Lewis and Wilson 1990). The alteration associated with mineralization is confined to narrow vein selvages of sericite-chlorite-pyrite±epidote. Sericitic alteration is dominant (A.-K. Appleby, written comm. 1996). The vein mineralogy, texture, and associated alteration are characteristic of epithermal mineralization in a low-sulphidation hydrothermal system (Appleby et al. 1995; White et al. 1995).

In summary, the epithermal gold-silver mineralization at Misima is attributed to coeval Pliocene extensional and transpressional tectonics, the unroofing of a metamorphic core complex, and mantle-derived potassic magmatism (A.-K. Appleby, written comm. 1996).

Petrography and Geochemistry of the Associated Potassic Rocks. The Miocene (8 Ma) Boiou microgranodiorite stocks and dykes vary texturally from porphyritic to even grained, with phenocrysts of plagioclase (< 5 mm long), quartz (~ 1 mm long), biotite (< 2 mm long), and amphibole (< 2 mm long) within a fine-grained groundmass consisting of quartz and feldspar (Lewis and Wilson 1990).

The Pliocene (3.5 Ma) lamprophyre dykes at Misima consist of up to 40 vol. % amphibole phenocrysts (~ 1 mm long) within a fine-grained groundmass dominated by plagioclase and secondary chlorite (A.-K. Appleby, written comm. 1996), and thus can be classified as spessartites as defined by Rock (1991). The spessartites near the Umuna Fault Zone are strongly carbonate altered (Appleby et al. 1995).

Geochemically, the microgranodiorites (Table 6.11) are characterized by evolved

Table 6.11. Major- and trace-element analyses of microgranodiorite and spessartites from Misima gold deposit, Papua New Guinea. Major elements are in wt %, and trace elements are in ppm. Fe_2O_3 (tot) = total iron calculated as ferric oxide. From unpublished data of Placer Pacific Ltd.

Province/deposit:	Misima	Misima	Misima	Misima	Misima
Location:	Papua New Guinea	Papua New Guinea	Papua New Guinea	Papua New Guinea	Papua New Guinea
Rock type:	Microgranodiorite	Spessartite	Spessartite	Spessartite	Spessartite
Tectonic setting:	Postcollisional arc	Postcollisional arc	Postcollisional arc	Postcollisional arc	Postcollisional arc
Reference:	A.-K. Appleby (written comm. 1996)	A.-K. Appleby (written comm. 1996)	A.-K. Appleby (written comm. 1996)	A.-K. Appleby (written comm. 1996)	A.-K. Appleby (written comm. 1996)
SiO_2	69.50	49.40	50.90	50.70	52.20
TiO_2	0.25	1.46	1.27	1.47	1.20
Al_2O_3	16.00	15.40	15.10	16.60	14.50
Fe_2O_3 (tot)	1.98	7.05	6.40	6.60	6.60
MnO	0.15	0.18	0.15	0.15	0.19
MgO	0.99	5.30	5.80	5.65	5.25
CaO	2.26	6.90	6.10	4.58	8.15
Na_2O	4.64	3.48	3.94	4.68	2.50
K_2O	3.06	1.21	2.38	2.34	4.20
P_2O_5	0.20	0.69	0.66	0.73	1.32
LOI	1.58	9.80	7.95	7.00	4.06
Total	100.61	100.87	100.65	100.50	100.17
mg#	54	64	68	66	65
V	40	180	170	170	160
Cr	130	70	180	60	150
Ni	<50	<50	100	<50	100
Rb	70	25	45	45	65
Sr	1040	160	320	260	1600
Y	6	14	11	15	13
Zr	100	160	150	160	220
Nb	10	15	10	15	15
Ba	930	80	760	940	1540
La	6	25	20	25	50
Ce	11	50	40	45	95
Th	n.a.	n.a.	n.a.	n.a.	n.a.
Ta	n.a.	n.a.	n.a.	n.a.	n.a.
Hf	5	6	6	6	8

compositions with high SiO_2 (< 69.5 wt %) contents, and very low TiO_2 (~ 0.25 wt %), Fe_2O_3 (i.e. ~ 1.98 wt %), and MgO (~ 0.99 wt %) concentrations (A.-K. Appleby, written comm. 1996). On a total alkalis versus silica plot (Le Maitre 1989), the rocks plot on the boundary between trachytes and dacites due to their high alkali contents (i.e. < 4.64 wt % Na_2O, < 3.06 wt % K_2O).

The spessartites (Table 6.11) have typical shoshonitic compositions (A.-K. Appleby, written comm. 1996), with relatively low SiO_2 (< 52.2 wt %) abundances, low but variable Al_2O_3 (14.5–16.6 wt %), high LILE (up to 4.2 wt % K_2O, up to 1540 ppm Ba, up to 1600 ppm Sr), relatively low LREE (< 50 ppm La, < 95 ppm Ce), and, except for TiO_2 and Nb, low HFSE (< 220 ppm Zr, ~ 6 ppm Hf) contents. The TiO_2 and Nb concentrations are slightly elevated (up to 1.47 wt % TiO_2, up to 15 ppm Nb). Their compositions are consistent with those of potassic igneous rocks from other postcollisional arc settings (see Sects 4.2, 6.5.1, 6.5.3).

Based on both structural considerations (A.-K. Appleby, written comm. 1996) and their geochemical fingerprints, the lamprophyres are interpreted to have been intruded into a postcollisional arc (Fig. 6.22).

6.5.3 Porgera Gold Deposit, Papua New Guinea

Introduction. The Miocene Porgera gold deposit in the highlands of Papua New Guinea (Fig. 6.23) is an example of modern gold mineralization hosted by volatile-rich potassic igneous rocks in a postcollisional arc setting. The Porgera gold mine was officially opened in 1990, and is one of the top gold producers in the world (Richards et al. 1991).

Regional Geology. The Porgera igneous complex is of middle to late Miocene age (Fleming et al. 1986), comprising a series of plugs, stocks, and dykes of potassic alkaline composition. Porgera is located in a highland mountainbelt (Fig. 6.23) which was formed by Late Miocene continent-island arc collision (Richards et al. 1991). The region is cut by the Lagaip Fault Zone, which trends west-northwest, and separates deformed and metamorphosed volcanosedimentary rocks to the northeast from unmetamorphosed Jurassic and Cretaceous sedimentary rocks to the southwest (Richards et al. 1991). These supracrustal rocks were affected by rapid uplift during the Early Pliocene following the Late Miocene collision event (Hill and Gleadow 1989). The intrusion of the Porgera igneous complex immediately predates the main stage of tectonism in the highlands (Richards et al. 1991).

Nature of Epithermal Gold Mineralization. The Porgera gold deposit contains about 410 tonnes Au and 890 tonnes Ag (Richards 1990a), distributed between a large, lower-grade ore zone (3.5 ppm Au, 9.9 ppm Ag) and a smaller high-grade zone (26.5 ppm Au, 22.1 ppm Ag). Gold mineralization at Porgera was episodic (Handley and Henry 1990), with several overlapping stages of mineralization from early, low-grade gold disseminations to late, high-grade epithermal gold veins (Richards et al. 1991). Mineralization was within 1 Ma of the time of magmatism —

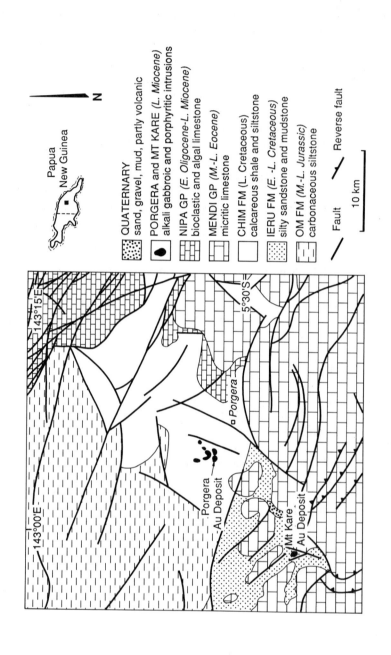

Fig. 6.23.   Geological overview of the Porgera and Mount Kare gold deposits, Papua New Guinea. Modified after Richards and Kerrich (1993).

which has been dated at 6.0 ± 0.3 Ma using $^{40}Ar/^{39}Ar$ ages of hornblende (Richards 1990a) — and appears to be genetically related to the alkalic Porgera intrusive complex (Richards 1992).

Richards (1990a) provides evidence for the evolution of a volatile phase during magma crystallization, suggesting that Au and other elements were partitioned into a magmatic fluid. These late-magmatic hydrothermal fluids probably caused the initial stage of gold enrichment (Richards 1990b), with mixing between this fluid and reduced groundwaters resulting in the deposition of base metals while Au was retained in solution as bisulphide complexes until precipitation at higher levels caused by either cooling, fluid mixing, or boiling (Richards 1990a). A late influx of fresh magma into the underlying magma chamber might have resulted in the emplacement of a late suite of feldspar porphyry dykes and the release of a final pulse of hydrothermal fluid (Richards 1990a). These fluids ascended late faults and precipitated the high-grade gold mineralization (Richards 1990a). The interpretation that Au may have been concentrated in a magmatic fluid phase is supported by the presence of hypersaline fluid inclusions and stable isotope compositions of mineralization-related minerals consistent with the involvement of magmatic volatiles (Richards 1992), and evidence for high halogen concentrations in mica phenocrysts of the magmatic host rocks (see Chap. 8). Although the highest gold contents in porphyry-type deposits are normally in the potassic alteration zone (Hollister 1975), no such alteration zone has been discovered at Porgera (Richards 1992).

Petrology and Geochemistry of the Potassic Host Rocks. The intrusions of the Porgera igneous complex are medium to coarse grained, porphyritic to ophitic, and consist of olivine, diopside, feldspar, hornblende, phlogopite, apatite, and magnetite (Richards 1990a; Richards and Ledlie 1993).

The potassic host rocks at Porgera are characterized by a primitive geochemistry (Table 6.12) with low SiO_2 (< 49.0 wt %), high Na_2O (up to 3.2 wt %) and high MgO (up to 12 wt %) contents, high mg# (up to 75), and high concentrations of mantle-compatible trace-elements (e.g. > 200 ppm V, up to 780 ppm Cr, up to 380 ppm Ni). The rocks have high LILE (e.g. ~ 650 ppm Sr, ~ 400 ppm Ba), moderate LREE (e.g. ~ 30 ppm La, ~ 60 ppm Ce), and, with the exception of Nb, low HFSE (< 1.20 wt % TiO_2, ~ 125 ppm Zr, ~ 3 ppm Hf) contents. Richards (1990b) interpreted the geochemistry of the Porgera intrusive suite in terms of within-plate magmatism resulting from localized melting in the subcontinental mantle. In terms of the available database, potassic igneous rocks from such within-plate settings have very high HFSE concentrations (Chap. 3). The Porgera volcanic rocks (Fig. 6.16), indeed, have very high Nb concentrations (~ 50 ppm), but the remaining HFSE have abundances which are far too low to be considered as reliable geochemical fingerprints of within-plate potassic volcanism (< 170 ppm Zr, < 19 ppm Y, < 3.6 ppm Hf). Potassic volcanic rocks from many mature postcollisional arc settings do tend to show a transition from a calc-alkaline or shoshonitic to a more alkaline geochemistry during the later stages of magmatism (e.g. Sect. 4.2; Müller et al. 1992a) which might apply at Porgera.

On balance, the relatively low HFSE (with the exception of Nb) contents of the

Table 6.12.   Major- and trace-element analyses of potassic igneous rocks from Porgera gold deposit, Papua New Guinea. Major elements are in wt %, and trace elements are in ppm. Fe_2O_3 (tot) = total iron calculated as ferric oxide. From Richards (1990a).

Province/deposit:	Porgera	Porgera
Location:	Papua New Guinea	Papua New Guinea
Rock type:	Basalt	Basalt
Tectonic Setting:	Postcollisional arc	Postcollisional arc
Reference:	Richards (1990a)	Richards (1990a)
SiO_2	48.25	45.08
TiO_2	0.88	1.19
Al_2O_3	15.50	13.03
Fe_2O_3 (tot)	7.07	9.47
MnO	0.16	0.17
MgO	5.73	12.23
CaO	9.61	11.41
Na_2O	3.27	2.14
K_2O	1.90	1.63
P_2O_5	0.34	0.64
LOI	7.58	3.05
Total	100.29	100.04
mg#	65	75
V	230	240
Cr	170	785
Ni	59	388
Rb	44	45
Sr	665	630
Y	15	16
Zr	127	120
Nb	59	49
Ba	415	330
La	31	32
Ce	59	65
Hf	2.8	2.9

Porgera intrusive suite implicate a subduction-related postcollisional arc setting rather than a within-plate tectonic setting. Moreover, a genetic model proposing that within-plate magmatism was responsible for the intrusion of the Porgera igneous complex at 6 Ma would require the uprise of a deep asthenospheric mantle plume with OIB-type geochemistry. Tectonic reconstructions of the area imply the presence of a oceanic slab subducting southwestward beneath the continental crust of the Papua New Guinea main island at this time (Cooper and Taylor 1987), which would have blocked the ascent of rising mantle plumes. Therefore, a postcollisional arc setting is suggested on the basis of both geological and geochemical data.

6.6   Synthesis of Direct Genetic Associations

As discussed above, there are spatial and probably genetic associations between copper-gold mineralization and high-K igneous suites in late oceanic arc, continental arc and postcollisional arc settings (cf. Müller and Groves 1993). It is apparent that Tertiary epithermal gold deposits (e.g. Ladolam, Emperor, Porgera; see Fig. 6.1) can be hosted by potassic igneous rocks in *late oceanic arcs*, commonly within or at the margin of a collapsed caldera structure (Anderson and Eaton 1990; Moyle et al. 1990). The Au-bearing sulphide mineralization in these settings is generally disseminated or occurs as quartz stockwork veining within the K-rich host rocks. The high-salinity fluid inclusions in these deposits suggest that the ore fluids were of magmatic origin (Kwak 1990; Moyle et al. 1990), and the mineralization and potassic magmatism are coeval. Older porphyry-style deposits at Goonumbla are from a similar tectonic setting, formed from high salinity fluids (Heithersay et al. 1990; Heithersay and Walshe 1995), and were coeval with high-K igneous rocks (e.g. Perkins et al. 1990a).

Most Cretaceous to Cenozoic epithermal (e.g. El Indio) and porphyry-type (e.g. Bajo de la Alumbrera) copper-gold deposits in the Chilean and Argentinian Andes are hosted by high-K calc-alkaline igneous rocks (e.g. Stults 1985; Tschischow 1989; Gröpper et al. 1991; Reyes 1991), and a direct genetic link between potassic magmatism and mineralization in this *continental arc* has been proposed.

A similar genetic association between epithermal gold mineralization and potassic igneous host rocks is also assumed at the Miocene Porgera and Pliocene Misima gold deposits, which occur in a *postcollisional arc*. Both hypersaline fluid inclusions and stable isotope data provide evidence for the involvement of magmatic volatiles in ore formation (Richards 1992), and gold mineralization occurred within 1 Ma of magmatism (Richards et al. 1991; A.-K. Appleby, written comm. 1996).

There is growing evidence that high-K igneous rocks also may be an important component of the setting of VMS deposits in late oceanic arcs (e.g. Fiji: Colley and Greenbaum 1980; Flin Flon, Manitoba, Canada: Syme and Bailes 1993; Stern et al. 1995) and postcollisional arc (e.g. western Tasmania: Crawford et al. 1992) settings. The role of the high-K rocks is not defined, and hence these associations are not discussed in detail in this book. One possibility is that the high-K igneous rocks mark tectonic settings significantly inboard of subduction where there is a greater chance of preservation of the deposits.

There is also evidence for an association between molybdenum mineralization and high-K igneous rocks, both in island arc settings (e.g. Polillo Island, Philippines: Knittel and Burton 1985) and from within-plate settings (e.g. Central City, Colorado: Rice et al. 1985).

7 Indirect Associations Between Lamprophyres and Gold-Copper Deposits

7.1 Introduction

Examples of ancient associations between gold mineralization and high-K rocks in postcollisional arc settings include the spatial and temporal associations between Archaean shoshonitic lamprophyres and mesothermal lode-gold deposits in the Superior Province, Canada (Wyman and Kerrich 1989a), including associations with very large gold deposits at Hollinger-McIntyre and Kerr Addison-Chesterville (Burrows and Spooner 1989; Spooner 1993), and in the Leonora-Laverton and New Celebration-Kambalda regions of the eastern Yilgarn Block, Western Australia (Perring et al. 1989a; Rock et al. 1989; Barley and Groves 1990; Taylor et al. 1994). These associations are discussed in Sections 7.5 and 7.4, respectively.

There are also somewhat similar associations between lamprophyres and gold-copper deposits in Proterozoic terrains, including those in the vicinity of the Goodall and Tom's Gully mines in the Pine Creek Geosyncline of the Northern Territory, Australia, as described in Sections 7.2 and 7.3, respectively. In this chapter, it is shown that although there is a strong spatial correlation between lamprophyres and gold deposits, this is an indirect rather than a genetic relationship.

7.2 Shoshonitic Lamprophyres with Elevated Gold Concentrations from the Goodall Gold Deposit, Northern Territory, Australia (Proterozoic)

7.2.1 Introduction

This section describes and discusses lamprophyre dykes from the Goodall gold deposit in the Pine Creek Inlier, Northern Territory, Australia, because of their spatial association with gold mineralization. The Goodall deposit is described in some detail here because much of the data was collected specifically for this study (cf. Müller

1993), but have not been published elsewhere. Goodall mine has been operated by Western Mining Corporation since 1988. Bulk reserves are estimated to be 27 tonnes of contained gold (D. Quick, pers. comm. 1992).

7.2.2 Regional Geology

The Pine Creek Geosyncline, which forms the major mineral province of the Northern Territory, consists of an Early Proterozoic (ca. 1900 Ma) metavolcanosedimentary sequence covering about 66000 km^2 between Darwin and Katherine (Needham et al. 1988; Needham and De Ross 1990). Three main geological units of the Pine Creek Geosyncline can be distinguished (Needham and De Ross 1990):

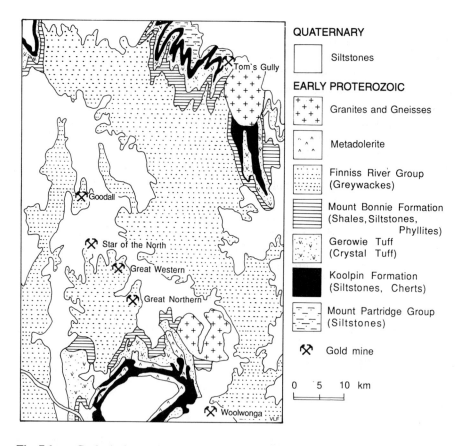

Fig. 7.1.   Geological overview of the Pine Creek Geosyncline, Northern Territory, Australia, showing the Goodall and Tom's Gully gold mines. The Mount Bonnie Formation, Gerowie Tuff, and Koolpin Formation comprise the South Alligator Group. Modified after Stuart-Smith et al. (1986).

- Early Proterozoic sedimentary rocks deposited in a shallow intra-cratonic geosyncline.
- Late Early Proterozoic rift-related felsic volcanic rocks.
- Subhorizontal platform sandstones of Middle Proterozoic age.

The Early Proterozoic sedimentary rocks consist of shales, siltstones, sandstones, conglomerates, and carbonates which were metamorphosed during the Top End Orogeny (ca. 1870-1690 Ma; Stuart-Smith et al. 1986; Page 1988). Geophysical modelling by Tucker et al. (1980) suggests that the basement is granitic throughout the whole region.

Regional metamorphism in the area decreases from upper amphibolite and granulite facies in the northeast, to lower greenschist facies in the centre, and increases to upper greenschist facies in the Rum Jungle area (Sheppard 1992). Most felsic and mafic intrusions in the area were emplaced during the Top End Orogeny (Needham et al. 1988). The earliest known granitoid intrusions are in the Nimbuwah domain to the east (ca. 1870 Ma) and in the Litchfield domain to the west (1850–1840 Ma; Needham et al. 1988). Lamprophyre intrusions are common in the central part of the Pine Creek Geosyncline (Stuart-Smith et al. 1986; Needham and Eupene 1990), and representative samples from the Mount Bundey area near Tom's Gully (Fig. 7.1) were dated via a Pb-Pb isochron at ca. 1831 ± 6 Ma (Sheppard and Taylor 1992).

Most economic deposits are restricted to the central part of the Pine Creek Geosyncline (Sheppard 1992). The area is notable as one of the world's largest and richest uranium provinces, as well as being a significant gold province within Australia (Needham and De Ross 1990). Despite the episodic nature of gold mineralization through Earth history, with production largely dominated by late Archaean and Mesozoic to Recent deposits, and only few discovered Proterozoic deposits (Barley and Groves 1992), there are several large gold producers in the Pine Creek Geosyncline (Sheppard 1992). Three styles of gold mineralization can be distinguished:

- Gold associated with uranium ores.
- Stratiform gold ores.
- Epithermal quartz-vein stockwork gold mineralization which is commonly accompanied by lamprophyre dykes.

The last-mentioned forms the most economically important style (e.g. Goodall and Woolwonga mines; Nicholson and Eupene 1990; Smolonogov and Marshall 1993). The heat source, and possibly the fluid and metal source, for much of the vein-type deposits were probably the Middle Proterozoic granites, spanning the period 1870–1765 Ma (Needham and Roarty 1980; Wall 1990; Sheppard 1992).

Table 7.1.   Microprobe (WDS) analyses of mica phenocrysts from lamprophyres at Goodall gold deposit, Northern Territory, Australia. Ox. form. = oxygen formula. Sample numbers refer to specimens held in the Museum of the Department of Geology and Geophysics, The University of Western Australia. Data from Müller (1993).

Sample no.:	119115	119115
wt %		
SiO_2	40.91	44.63
TiO_2	3.12	4.44
Al_2O_3	27.68	29.74
FeO (tot)	9.93	3.82
Cr_2O_3	0.02	0.02
MnO	0.04	0.04
MgO	6.55	3.33
NiO	0.06	0.05
BaO	0.29	0.31
CaO	0.03	0.03
SrO	0.12	0.10
K_2O	6.78	8.82
Na_2O	0.10	0.11
Cl	0.02	0.02
F	0.35	0.47
Total	96.00	96.24
mg#	60	67
Ox. form.	22	22
Atoms		
Si	5.861	6.037
Ti	0.323	0.451
Al	4.673	4.739
Fe	1.189	0.431
Cr	0.002	-
Mn	0.001	0.003
Mg	1.398	0.671
Ni	0.001	0.005
Ba	0.016	0.016
Ca	-	0.005
Sr	-	0.006
K	1.239	1.521
Na	0.028	0.030
Cl	-	-
F	-	-
Total	14.731	13.915

7.2.3 Nature of Mesothermal Gold Mineralization

The Goodall gold deposit is hosted by folded Early Proterozoic greywackes, siltstones, and shales of the Finniss River Group (Fig. 7.1) in the central part of the Pine Creek Geosyncline. Low-grade metamorphosed and folded sedimentary rocks are discordantly cut by lamprophyre dykes striking north-northwest. The sediment-hosted ore shoots also strike north-northwest. The thickness of dykes varies from 10 to 50 cm, and they are exposed in both the open pit and several diamond-drill holes. The lamprophyres appear to be broadly parallel to fold axial planes, and they either predate gold mineralization or have been intruded synchronously with it. All dykes are affected by hydrothermal alteration, which has produced secondary sericitization, and several dykes in the open pit are mineralized. Most dyke rocks analyzed in this study were sampled from exploration drill holes, with one sample collected in the open pit.

Sulphide mineralization at Goodall occurs mainly as epigenetic quartz stockwork veining in bleached sericitized siltstones, and rarely as disseminated assemblages, and it consists mainly of arsenopyrite, pyrite, and chalcopyrite. Native gold appears to be entirely vein-related, occurring as visible gold accompanied by arsenopyrite.

7.2.4 Mineralogy of the Lamprophyres

Most Goodall lamprophyres are characterized by cognate mica phenocrysts in a groundmass of feldspar and quartz. One sample contains amphibole phenocrysts which are completely altered to chlorite. Only their typical shapes allow them to be classified as former amphiboles. Two generations of mica phenocrysts with different sizes can be distinguished: a phenocryst phase with large phenocrysts (up to 4 mm) and a groundmass phase with small, elongated mica flakes (~ 1 mm). The rocks have been hydrothermally altered after or during dyke emplacement, with chloritization of the mica and amphibole phenocrysts (Table 7.1), and sericitization of groundmass feldspars.

7.2.5 Geochemistry of the Lamprophyres

The whole-rock major- and trace-element geochemistry of seven lamprophyres from Goodall gold mine is shown in Table 7.2. The investigated lamprophyres have andesitic compositions (51.4–63.1 wt % SiO_2) with low TiO_2 (< 0.8 wt %), high but variable Al_2O_3 (14.1–18.4 wt %), and high K_2O (> 2.3 wt %) contents. The K enrichment is probably due to secondary sericitization. The lamprophyres are characterized by extremely low Na_2O (< 0.25 wt %) and very low CaO (< 0.89 wt %) contents, which were probably caused by secondary alteration processes. The use of the K_2O versus SiO_2 biaxial plot of Peccerillo and Taylor (1976a), in order to determine the shoshonitic geochemistry of the samples, is flawed due to the mobilization of alkali elements during alteration. However, all samples plot in the shoshonite fields

Table 7.2. Major- and trace-element analyses of lamprophyres from Goodall gold deposit, Australia. Major elements are in wt %, trace elements are in ppm, and precious metals are in ppb. Fe_2O_3 (tot) = total iron calculated as ferric oxide. Precious-metal detection limits are: Au, Pt = 5 ppb, Pd = 1 ppb. Sample numbers refer to specimens held in the Museum of the Department of Geology and Geophysics, The University of Western Australia. Data from Müller (1993).

Sample no.:	119111	119112	119113	119114	119115	119116	119117
SiO_2	51.40	54.80	63.10	54.10	56.00	60.10	54.80
TiO_2	0.80	0.85	0.63	0.85	0.87	0.74	0.81
Al_2O_3	15.60	14.13	18.46	16.66	17.06	16.25	16.02
Fe_2O_3 (tot)	16.07	10.47	5.18	13.93	9.58	9.20	14.28
MnO	0.31	0.06	0.02	0.04	0.05	0.02	0.04
MgO	5.01	9.29	2.95	5.72	7.18	4.96	5.72
CaO	0.37	0.89	0.17	0.26	0.24	0.03	0.27
Na_2O	0.16	0.25	0.15	0.12	0.16	0.16	0.13
K_2O	3.57	2.31	5.27	3.68	3.10	3.80	3.13
P_2O_5	0.21	0.67	0.23	0.22	0.15	0.22	0.20
LOI	6.45	6.39	4.02	4.46	5.72	4.62	4.70
Total	99.96	100.14	100.13	100.02	100.11	100.10	100.07
mg#	42	67	57	49	63	55	48
V	164	227	107	185	166	206	167
Ni	38	68	28	12	59	81	42
Cu	171	67	20	265	20	20	20
Zn	109	92	60	62	105	78	48
As	n.a.	n.a.	32	250	34	75	61
Rb	164	89	302	139	140	209	147
Sr	16	71	39	9	11	11	11
Y	18	26	30	22	20	20	24
Zr	148	258	206	154	166	139	149
Nb	12	10	14	9	5	6	6
Sb	n.a.	n.a.	2.4	4.4	3.9	11.0	2.1
Ba	860	514	616	274	142	353	249
La	32.9	43.6	49.8	37.0	34.3	27.9	35.0
Ce	63.0	80.0	86.0	69.0	64.0	49.0	66.0
Yb	1.6	1.6	2.8	1.8	1.9	1.5	1.9
Hf	3.2	5.5	5.2	3.4	3.6	3.0	3.5
Ta	0.7	0.6	0.8	0.5	0.4	0.4	0.4
W	n.a.	n.a.	4.5	14.0	2.0	3.0	4.5
Th	15	17	18	14	13	8	14
Pd	n.a.	n.a.	< 1	< 1	<1	< 1	< 1
Pt	n.a.	n.a.	< 5	< 5	< 5	< 5	< 5
Au	59	5	28	54	5	5	5

of the Ce/Yb versus Ta/Yb and the Th/Yb versus Ta/Yb biaxial plots (Fig. 7.2). Since these elements are considered to be essentially immobile during secondary alteration processes (Pearce and Cann 1973; Pearce 1982), the plots allow the classification of the dykes as shoshonitic lamprophyres. For comparison, the lamprophyres from the Mount Bundey area, located about 40 km northeast of Goodall gold mine (Sheppard and Taylor 1992), have also been plotted on Figure 7.2. They form a different dyke swarm with a distinctive geochemistry (see Sect. 7.3).

The lamprophyres range from fractionated to relatively primitive (mg# of 42-67), with high V (~ 160 ppm) and moderate Ni (~ 40 ppm) concentrations. The low Ba

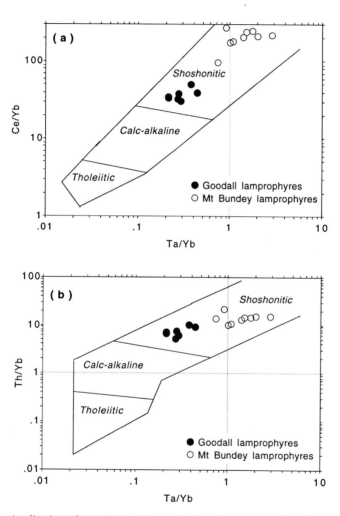

Fig. 7.2.   Application of geochemical discrimination diagrams based on immobile trace-elements (after Pearce, 1982) in order to illustrate the shoshonitic character of the highly altered lamprophyres from Goodall, Northern Territory. (a) (Ce/Yb) versus (Ta/Yb) plot. (b) (Th/Yb) versus (Ta/Yb) plot. Data for the Goodall lamprophyres are from Müller (1993) and those for lamprophyres from the Mount Bundey area, about 40 km northeast, are from Sheppard and Taylor (1992).

and Sr concentrations and the Ba/Nb ratios of < 71, which are relatively low in comparison with mica-phyric shoshonitic magmas from other localities (e.g. Chap. 3), are not primary features and are likely to have been caused by mobilization of these elements during secondary alteration processes. The original rocks were probably characterized by very high enrichments of LILE, a common feature of unaltered shoshonitic rocks.

The investigated lamprophyre dykes have a different geochemistry from those of the Mount Bundey area (Sheppard and Taylor 1992). In comparison, the Goodall lamprophyres are characterized by higher SiO_2 (> 51 wt %) and Al_2O_3 (> 14 wt %), and much lower P_2O_5 contents (< 0.6 wt %) than those from the Mount Bundey suite (generally < 50 wt %, < 13 wt %, > 0.7 wt %, respectively). The Goodall lamprophyres are also strongly affected by secondary alteration, as reflected in their extremely low Na_2O and CaO contents mentioned above. The suites differ in their trace-element geochemistry, with much lower LILE (< 302 ppm Rb, < 71 ppm Sr), lower LREE (< 49 ppm La, < 86 ppm Ce), and lower HFSE (< 0.87 wt % TiO_2, < 258 ppm Zr, < 14 ppm Nb, < 5.5 ppm Hf) concentrations for the Goodall lamprophyres. The Mount Bundey dykes, as described by Sheppard and Taylor (1992), contain very high LILE (up to 355 ppm Rb, up to 3635 ppm Sr), very high LREE (up to 340 ppm La, up to 750 ppm Ce), and higher HFSE (up to 2.1 wt % TiO_2, up to 932 ppm Zr, up to 75 ppm Nb, up to 21 ppm Hf) contents which are typical for potassic igneous rocks, such as lamprophyres, generated in a within-plate tectonic

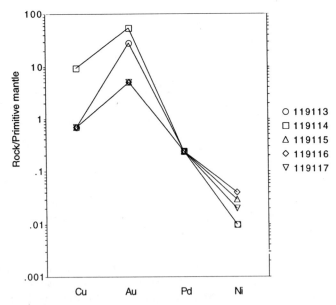

Fig. 7.3.   Abundances of chalcophile elements in lamprophyres from Goodall gold deposit, Northern Territory relative to the primitive mantle. Normalizing factors after Brügmann et al. (1987). Data from Müller (1993).

Table 7.3.   Correlation matrix for precious metals (Au, Pd, Pt), Cu, and gold pathfinder elements (As, Sb, W) of lamprophyres from the Goodall gold deposit, Australia. Data from Müller (1993).

	Cu	Au	Pt	Pd	As	Sb	W
Cu	1						
Au	0.889	1					
Pt	-	-	1				
Pd	-	-	-	1			
As	0.98	0.817	-	-	1		
Sb	-0.056	-0.228	-	-	0.084	1	
W	0.973	0.927	-	-	0.956	-0.151	1

setting (Müller et al. 1993a). However, the use of the geochemical discrimination diagrams (see Chap. 3) in order to determine the tectonic setting of the Goodall lamprophyres is not appropriate because of the strong hydrothermal alteration.

7.2.6 Direct or Indirect Link Between Potassic Lamprophyres and Mineralization

A possible genetic link between shoshonitic lamprophyres and mesothermal gold mineralization (as discussed above) has been proposed by Rock and Groves (1988a, 1988b). The intrusion of lamprophyre magmas into the crust is capable of promoting hydrothermal circulation and initiating partial melting, thereby generating felsic igneous rocks which are commonly associated with gold deposits (e.g. Sheppard 1992). During crystallization, lamprophyres can generate S- and CO_2-rich fluids, analogous to those thought responsible for the genesis of mesothermal gold deposits (Rock 1991).

Some lamprophyres from the Goodall deposit indeed show enrichments in Au of up to 59 ppb (Table 7.2). The elevated Au values are, however, unlike those of the potassic rocks from the Karinya Syncline (Chap. 5) in that they are *decoupled* from Cu and Pd peaks in primitive mantle-normalized distribution plots (Fig. 7.3; after Brügmann et al. 1987), suggesting that the anomalous Au contents are secondary features (Wyman and Kerrich 1989a). The high Au enrichment compared to other chalcophile elements such as Cu and Pd is reflected in the high Au/Pd (up to 54) and Au/Cu ratios (up to 1.4), compared to primitive mantle ratios of 0.25 and 0.036, respectively (Brügmann et al. 1987). Hydrothermal mineralizing fluids which overprinted the dykes *after* emplacement are thought to be responsible for the elevated gold concentrations in the dykes. This interpretation is consistent with the high correlation between Au and its pathfinder elements (e.g. As, W), as shown in Table 7.3. In comparison to the Truro lamprophyres from South Australia (see Sect. 5.3.), the lamprophyres at Goodall are situated in a major gold-mining area and were intensively altered (secondary sericitization) by hydrothermal fluids after emplacement.

7.3   Shoshonitic Lamprophyres from the Tom's Gully Gold Deposit, Northern Territory, Australia (Proterozoic)

7.3.1   Introduction

No direct genetic associations between potassic igneous rocks from within-plate tectonic settings and economic gold or base-metal mineralization have been reported to date. However, indirect associations, where both gold mineralization and potassic lamprophyre dykes were emplaced along major fault zones, are apparent in some areas, as discussed in Section 7.2 for the Goodall district.

A further example of a probable within-plate association is the Tom's Gully gold deposit in the Mount Bundey area, Northern Territory, Australia.

7.3.2   Regional Geology

The Mount Bundey area is located in the northern part of the Pine Creek Geosyncline (Fig. 7.1). As described in Section 7.2, the Pine Creek Geosyncline consists of Early Proterozoic metasedimentary and minor metavolcanic rocks. Rocks of the Pine Creek Geosyncline unconformably overlie several Archaean granitic and gneissic complexes thought to form a continuous basement (Needham et al. 1988). The main period of deformation and metamorphism is dated at 1885–1860 Ma (Page 1988).

The Mount Bundey pluton, covering about 80 km^2, consists of post-tectonic syenitic and granitic rocks which intrude shales and siltstones of the Early Proterozoic South Alligator Group (Fig. 7.1). The lamprophyre dykes are restricted to within 10 km of the Mount Bundey pluton, which is probably coeval with mineralization (Sheppard 1992). The dykes intrude the syenites and I-type granites of the pluton, and postdate regional deformation and metamorphism (Sheppard 1992, 1995). Both the lamprophyres and the Mount Bundey pluton are dated via conventional U-Pb in zircon methods at 1831 ± 6 Ma (Sheppard 1992, 1995).

7.3.3   Nature of Mesothermal Gold Mineralization

The Tom's Gully gold deposit is situated in the thermal aureole of the Mount Bundey pluton. There are two sulphidic ore-shoots in a single, fault-controlled, shallowly-dipping quartz reef (Sheppard 1992). On average, the reef is 1.0–1.5 m thick, but it pinches and swells between 0 and 2.4 m. Ore was probably deposited at Tom's Gully during wrench shearing associated with emplacement of the granitic rocks (Sheppard 1992). The sulphidic ore-shoots consist mainly of pyrite-arsenopyrite±loellingite±gold (Sheppard 1992), which are hosted by graphitic siltstones and shales of the Wildman Siltstone Unit (Sheppard 1992). Loellingite is commonly replaced by arsenopyrite. Visible gold commonly occurs as blebs of electrum within arsenopyrite. Minor wallrock alteration of the metasedimentary host rocks is expressed by the oxidation of graphitic pelites and formation of secondary sericite and potassic feldspar (Sheppard 1992).

7.3.4 Petrology and Geochemistry of the Lamprophyres

The lamprophyre dykes are 0.5–3.0 m thick and have chilled margins. Two petrographic types can be distinguished (Sheppard and Taylor 1992): one olivine-phlogopite-diopside-phyric and the other olivine-amphibole-diopside-phyric. Both types contain apatite microphenocrysts and their groundmass is dominated by orthoclase.

Geochemically, the lamprophyres are characterized by very high F (up to 4900 ppm), very high P_2O_5 (~ 2.0 wt %), and K_2O (up to 7.4 wt %) contents (Table 7.4), and resulting high K_2O/Na_2O ratios (> 2; Sheppard and Taylor 1992). The Mount Bundey lamprophyres have a primitive geochemistry (Sheppard and Taylor 1992) with relatively high mg# (63–66) and high mantle-compatible element concentrations (e.g. > 107 ppm V, > 205 ppm Cr, > 157 ppm Ni). Their high LILE (e.g. up to 3635 ppm Sr, up to 5101 ppm Ba), high LREE (e.g. ~ 220 ppm La, ~ 500 ppm Ce) and very high HFSE (~ 2 wt % TiO_2, ~ 750 ppm Zr, ~ 60 ppm Nb, ~ 17 ppm Hf) contents are consistent with potassic igneous rocks emplaced in a within-plate tectonic setting (cf. Figs 6.3, 6.4).

The coincidence of lamprophyres, syenites, and granites is common in many localities worldwide (Rock 1991). Fractional crystallization of lamprophyric melts has been demonstrated to produce syenitic magmas (McDonald et al. 1986; Leat et al. 1988), and granites associated with lamprophyres are interpreted to be generated by crustal assimilation triggered by the interaction of mantle-derived hot and volatile-rich lamprophyric melts with the lower crust (McDonald et al. 1986). It is possible that the conduits for these magmas coincide with those for deeply sourced auriferous fluids.

7.3.5 Indirect Link Between Lamprophyres and Gold Mineralization

The lamprophyres, syenite and granite at Mount Bundey define a Pb-Pb isochron age of 1831 ± 6 Ma, suggesting that the three rock types form a co-magmatic suite (Sheppard and Taylor 1992) in which the syenite represents the fractionated product of lamprophyric magmatism. Although the gold mineralization at Tom's Gully is bracketed in time by the lamprophyre dykes, the lamprophyres do not have intrinsically high Au contents and no direct genetic association is evident (Sheppard 1992). The shoshonite-gold association at Tom's Gully is interpreted to be indirect, representing only similar structural controls on both dyke intrusion and mineralization, as at Goodall in the same terrane.

Table 7.4. Major- and trace-element analyses of lamprophyres from the Mount Bundey district, Northern Territory, Australia. Major elements are in wt %, and trace elements are in ppm. Fe_2O_3 (tot) = total iron calculated as ferric oxide. Data from Sheppard and Taylor (1992).

Province/deposit:	Mount Bundey	Mount Bundey
Location:	Northern Territory, Australia	Northern Territory, Australia
Rock type:	Lamprophyre	Lamprophyre
Tectonic setting:	Within-plate	Within-plate
Reference:	Sheppard and Taylor (1992)	Sheppard and Taylor (1992)
SiO_2	46.73	47.22
TiO_2	2.13	1.92
Al_2O_3	11.88	11.30
Fe_2O_3 (tot)	11.02	10.30
MnO	0.14	0.13
MgO	8.12	8.65
CaO	7.78	7.35
Na_2O	2.34	2.18
K_2O	5.82	6.49
P_2O_5	2.08	2.18
LOI	2.07	2.01
Total	100.11	99.73
mg#	63	66
V	154	144
Cr	205	262
Ni	200	237
Rb	162	181
Sr	3136	3635
Y	40	37
Zr	722	866
Nb	59	70
Ba	5023	5101
La	220	228
Ce	483	514
Hf	16	19

7.4 Shoshonitic Lamprophyres from the Eastern Goldfields, Yilgarn Block, Western Australia (Archaean)

7.4.1 Introduction

In Australia, the ca. 2700 Ma granitoid-greenstone terrains of the Yilgarn Block are the most intensely mineralized with world-class gold deposits (Groves et al. 1994). Widespread shoshonitic lamprophyre dyke-swarms also represent a significant contribution to magmatism in the Yilgarn Block at ca. 2680–2660 Ma (Rock et al. 1988b). Along with contemporaneous swarms from the Superior Province, Canada, and the Limpopo Belt, Zimbabwe, they may represent a global Archaean mantle event (Rock et al. 1988b). Many of these lamprophyre dykes have been misclassified as "diorites", "diabases" and "trachyandesites" in the past (Rock 1991).

7.4.2 Regional Geology

The Yilgarn Block comprises high-grade gneiss and supracrustal rocks in the west and granitoid-greenstone terrains in its central and eastern segments (Groves et al. 1994). The craton has traditionally been subdivided into four main subprovinces: the Western Gneiss Terrain, and the Murchison, Southern Cross, and Eastern Goldfields Provinces (Gee et al. 1981; Fig. 7.4). Myers (1993) redefines the craton as a number of geologically distinct superterranes. However, the province terminology is well recognized in the literature so it is used here.

The Western Gneiss Terrain contains the oldest Archaean crust recognized to date in Australia (Groves et al. 1994). The granitoid-greenstone terrains of the Murchison, Southern Cross, and Eastern Goldfields Provinces have a common history of granitoid intrusion, deformation, and metamorphism from ca. 2680 to 2630 Ma, although the trend of major fault and shear zones varies from one province to another (Groves et al. 1994). The Murchison Province, for example, is dominated by northeast-trending shear zones and greenstone belts, whereas most shear zones and greenstone belts trend north-northwest to northwest in the Eastern Goldfields Province.

Although there are only preliminary data, there appear to be two major supracrustal sequences in the greenstone belts. Older sequences, which are dominated by tholeiitic to high-magnesium basalts, contain abundant banded iron-formations (BIF). These sequences are dominant in the Murchison and Southern Cross Provinces, with only local occurrences in the Eastern Goldfields Province. In contrast, younger supracrustal sequences, dated at ca. 2700 Ma, dominate the Norseman-Wiluna Belt in the Eastern Goldfields Province, but are more restricted in the Murchison and Southern Cross Provinces (Groves et al. 1994). These sequences contain virtually no BIF, but there are thick sequences of komatiites and discrete felsic volcanic centres (e.g. Hallberg 1985). The volcanic rocks are commonly overlain by clastic sedimentary rocks in restricted structural basins. The ca. 2700 Ma volcanic sequences resemble those in modern subduction arcs (e.g. Barley et al. 1989). However, the volcanic rocks were apparently erupted through older continental crust because they commonly contain xenocrystic zircons older than 3000 Ma (Groves et al. 1994).

7.4.3 Nature of Mesothermal Gold Mineralization

The Western Gneiss Terrain, like most high-grade gneiss terrains worldwide, contains only minor mineralization. In contrast, the granitoid-greenstone terrains are exceptionally well mineralized with widespread lode-gold and komatiite-hosted nickel-copper, and more restricted copper-zinc VMS deposits (Groves et al. 1994).

With an output of about 2037 tonnes of gold to 1987, the Yilgarn Block has produced almost half of Australia's gold production from lode deposits of about 4375
tonnes to 1987 (Groves et al. 1994). The Norseman-Wiluna Belt is the most highly
mineralized, followed by the Murchison Province, the Southern Cross Province, and
the remainder of the Eastern Goldfields Province. About half of the total gold production has come from the Golden Mile at Kalgoorlie. Although the physical ap-
pearance of the lode-gold deposits varies greatly, due to differences in structural
style, host rock, and mineralogy of alteration assemblages, the deposits appear to
represent a coherent genetic group (Groves 1993). Basically, they comprise structurally controlled gold±silver±arsenic±tellurium±antimony±tungsten deposits, associ-

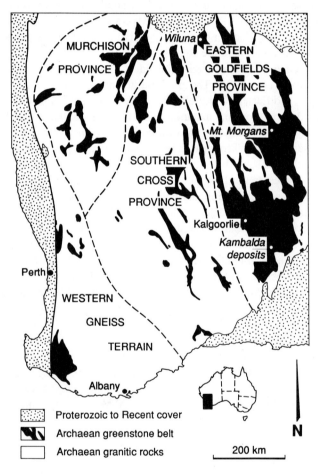

Fig. 7.4.   Geological overview of the Yilgarn Block, Western Australia, showing the Eastern Goldfields Province and the Kambalda, Mount Morgans and Wiluna gold deposits, which
are spatially associated with shoshonitic lamprophyres. After Gee et al. (1981).

ated with metasomatic zones representing $K\pm CO_2\pm Na\pm Ca$ addition, in a variety of ultramafic, mafic and felsic igneous rocks, and Fe-rich sedimentary rocks in greenstone belts of sub-greenschist to lower granulite-facies grade (normally greenschist-amphibolite; Groves 1993; Solomon and Groves 1994).

7.4.4 Lamprophyres and Their Association with Mineralization

Shoshonitic lamprophyres have been reported from the Eastern Goldfields (e.g. Kambalda and Leonora-Laverton areas) and Murchison Provinces (Rock et al. 1988b). Most of these lamprophyres (see Fig. 7.4) occur in the highly mineralized greenstone belts of the Eastern Goldfields Province (Hallberg 1985), in particular in association with the mesothermal gold deposits at Wiluna (Hagemann et al. 1992), Mount Morgan (Vielreicher et al. 1994) and Kambalda (Perring 1988). Mutually cross-cutting relationships between these dykes and mineralized quartz veins at several mines suggest that shoshonitic magmatism overlapped with, although it was commonly earlier than, the period of gold mineralization (Hallberg 1985; Barley and Groves 1990; Taylor et al. 1994). Both lamprophyres and gold mineralization appear to be spatially and genetically linked to subhorizontal oblique compression, and occur along major shear zones which are interpreted to have extended to the mantle (Perring et al. 1989a; Wyman and Kerrich 1988, 1989a), thus providing favourable conduits for both lamprophyric magmas and deeply sourced mineralizing fluids.

7.4.5 Petrology and Geochemistry of the Lamprophyres

The shoshonitic lamprophyres comprise amphibole-phyric spessartites and mica-phyric kersantites (Perring 1988; Rock et al. 1988b). Spessartites normally consist of euhedral, zoned hornblende phenocrysts set in a groundmass of plagioclase with accessory apatite and titanite. Kersantites are characterized by battlemented phlogopite phenocrysts in a groundmass containing plagioclase and carbonate with accessory pyrite, apatite, and zircon. Felsic ocelli are common in the dykes and are composed of arborescent plagioclase (larger crystals show chessboard albite twinning) with minor carbonate (Perring 1988). Some kersantites also contain quartz xenocrysts (Perring 1988), which is indicative of volatile-driven rapid uprise of the lamprophyric magma (cf. Rock 1991).

The shoshonitic lamprophyres of the Eastern Goldfields Province are altered to various degrees. Fresh samples are characterized geochemically by relatively primitive compositions with relatively high, but variable, mg# (up to 67) and high concentrations of mantle-compatible elements (e.g. > 100 ppm V, > 300 ppm Cr, > 100 ppm Ni; Taylor et al. 1994). They normally have high LILE (e.g. up to 740 ppm Sr, up to 1300 ppm Ba), low LREE (e.g. ~ 40 ppm La, ~ 75 ppm Ce), and very low HFSE (e.g. < 0.7 wt % TiO_2, < 160 ppm Zr, < 8 ppm Nb, < 5 ppm Hf) contents. Representative data of mineralogically fresh samples are presented in Table 7.5. It should

Table 7.5.   Major- and trace-element analyses of lamprophyres from the Eastern Goldfields Province, Yilgarn Block, Western Australia. Major elements are in wt %, and trace elements are in ppm. Fe_2O_3 (tot) = total iron calculated as ferric oxide. Data from Taylor et al. (1994).

Province/deposit:	Eastern Goldfields, Yilgarn Block	Eastern Goldfields, Yilgarn Block
Location:	Western Australia	Western Australia
Rock type:	Lamprophyre	Lamprophyre
Tectonic setting:	Postcollisional arc	Postcollisional arc
Reference:	Taylor et al. (1994)	Taylor et al. (1994)
SiO_2	60.39	55.99
TiO_2	0.55	0.69
Al_2O_3	14.56	12.61
Fe_2O_3 (tot)	5.96	7.52
MnO	0.09	0.12
MgO	5.17	8.46
CaO	6.24	5.74
Na_2O	3.73	3.82
K_2O	1.84	3.10
P_2O_5	0.22	0.36
LOI	1.37	2.47
Total	100.12	100.88
mg#	67	52
V	116	135
Cr	323	684
Ni	101	217
Rb	29	75
Sr	589	738
Y	14	23
Zr	131	154
Nb	5	8
Ba	359	1278
La	38	35
Ce	72	77
Hf	3.1	4.4

be noted that the Au content of most Yilgarn lamprophyres is only elevated in those samples which were collected from the vicinity of mesothermal gold deposits (Fig. 7.5). The gold values are decoupled from those of other chalcophile elements in primitive mantle-normalized distribution plots (Fig. 7.5b), suggesting that they are secondary features. This is consistent with the poor correlation between Au and the magmatic elements Cu, Pt, and Pd (Table 7.6). The significance of the gold contents of the rocks in terms of an indirect or genetic association between lamprophyres and gold mineralization is discussed in Section 7.6.

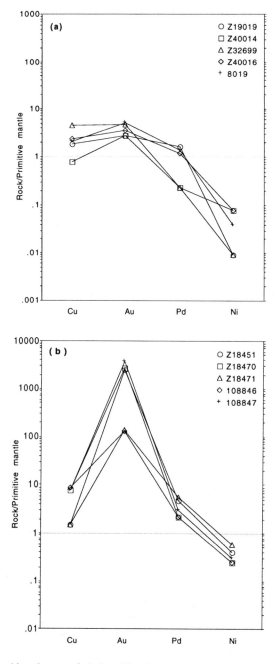

Fig. 7.5. Abundances of chalcophile elements in lamprophyres from the Yilgarn Block relative to the primitive mantle. (a) Data for lamprophyres distal to gold deposits. (b) Data for lamprophyres from the vicinity of gold deposits. Normalizing factors after Brügmann et al. (1987). Data from Taylor et al. (1994).

Table 7.6. Correlation matrix for precious metals (Au, Pd, Pt), Cu, and gold pathfinder elements (As, Sb, W) of lamprophyres from the Yilgarn Block. (a) Data for samples distal to gold deposits. (b) Data for samples proximal to gold deposits. Data from Taylor et al. (1994).

(a)

	Cu	Au	Pt	Pd	As	Sb	W
Cu	1						
Au	0.713	1					
Pt	0.312	0.409	1				
Pd	0.258	0.512	-0.091	1			
As	-0.399	0.023	-0.202	-0.434	1		
Sb	0.102	-0.123	-0.674	0.634	-0.525	1	
W	-0.304	-0.245	-0.627	0.682	-0.365	0.877	1

(b)

	Cu	Au	Pt	Pd	As	Sb	W
Cu	1						
Au	0.228	1					
Pt	-0.563	0.082	1				
Pd	-0.342	-0.420	0.631	1			
As	-0.040	-0.067	0.154	-0.424	1		
Sb	0.286	0.631	0.012	0.182	-0.780	1	
W	0.092	0.791	-0.102	-0.173	-0.628	0.878	1

7.5   Shoshonitic Lamprophyres from the Superior Province, Canada (Archaean)

7.5.1   Introduction

The Superior Province comprises a collage of granitoid-greenstone belts with intervening belts of metasedimentary rocks and tonalitic gneisses which are separated by major structural discontinuities such as shear zones and transcurrent fault systems (Wyman 1990). The greenstone belts show a similar age range to those of the Yilgarn Block described in Section 7.4, but there is much less evidence for eruption through continental crust (e.g. Wyman 1990). Alkaline igneous rocks are only a volumetrically minor component of the Archaean Superior Province of Canada but, as in Western Australia, share important spatial and temporal distributions with lode-gold deposits (Wyman and Kerrich 1988). The alkalic magmatism comprises both shoshonitic lamprophyres and K-rich plutonic rocks which are sited along major structures and are late in the greenstone belt development (Wyman and Kerrich 1988).

7.5.2   Nature of Mesothermal Gold Mineralization

Superior Province lode-gold deposits systematically occur as late kinematic features, generally in proximity to crustal-scale structures called "breaks" (Wyman and Kerrich 1988). Similar to the lode-gold deposits in the Yilgarn Block, those in the Superior Province, also comprise mainly structurally controlled gold±silver± arsenic±tellurium±antimony±tungsten deposits, which are associated with metasomatic zones representing $K\pm CO_2\pm Na\pm Ca$ addition. The mesothermal lode-gold deposits are associated with a variety of ultramafic, mafic, and felsic igneous rocks and Fe-rich sedimentary rocks in greenstone belts of greenschist- to amphibolite-facies grade. An excellent summary is given by Colvine (1989).

7.5.3   Lamprophyres and Their Association with Mineralization

Shoshonitic lamprophyre dykes are abundant in areas such as Kirkland Lake, where several small syenite plutons are also prominent (Rowins et al. 1993), and in the Hemlo region, which is transected by regional-scale shear zones (Wyman and Kerrich 1988). Both mesothermal gold mineralization and lamprophyre emplacement young to the south from Red Lake (ca. 2700 Ma) to the southern Abitibi Belt (ca. 2673 Ma; Wyman and Kerrich 1989a). The transgressive nature of these events is compatible with a series of accretionary events which resulted in generation of the greenstone belts in the Late Archaean (Wyman and Kerrich 1989a), implying that these processes were linked to a postcollisional arc setting.

The shoshonitic lamprophyres are typically restricted to granitoid-greenstone subprovinces and their tectonic boundaries with metasedimentary terranes (Wyman 1990), a geological setting which they share with their Western Australian counterparts (Groves et al. 1994). The Superior Province lamprophyres are dated at ca. 2690–2675 Ma, and include minette, kersantite, and vogesite dykes (Wyman and Kerrich 1989a). The steeply dipping dykes contain phenocrysts of augite, phlogopite and/or hornblende in a primary carbonate-bearing, feldspathic groundmass (Wyman 1990), and their emplacement is interpreted to be spatially and genetically linked to subduction-like underthrusting within a largely transpressive tectonic regime (e.g. Wyman and Kerrich 1989a). The lamprophyre-gold spatial association is restricted to the final period of stabilization of the Superior Province (Wyman 1990), being particularly well developed near major fault systems, as exemplified by the Kirkland Lake area (Fig. 7.6) and Abitibi Belt (Jensen 1978; Toogood and Hodgson 1985; Wyman 1990; Rowins et al. 1993). In the Kirkland Lake area, both lamprophyres and gold mineralization are also spatially associated with high-K syenitic plutons (Fig. 7.6; Mortensen 1993) which are dated at 2680 ± 1 Ma using conventional U-Pb in zircon techniques (Rowins et al. 1993). The syenitic plutons are believed to have been generated by fractional crystallization of a mantle-derived lamprophyric magma (Rowins et al. 1993).

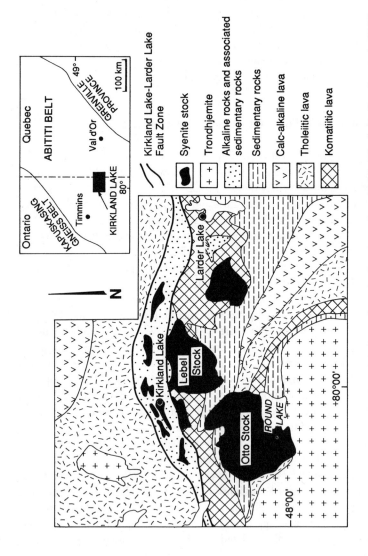

Fig. 7.6. Geological overview of the Kirkland Lake area, Superior Province, Canada. Modified after Rowins et al. (1993).

Table 7.7. Major- and trace-element analyses of lamprophyres from the Superior Province, Canada. Major elements are in wt %, and trace elements are in ppm. Fe_2O_3 (tot) = total iron calculated as ferric oxide. Data from Wyman (1990).

Province/deposit:	Superior	Superior
Location:	Canada	Canada
Rock type:	Lamprophyre	Lamprophyre
Tectonic setting:	Postcollisional arc	Postcollisional arc
Reference:	Wyman (1990)	Wyman (1990)
SiO_2	48.60	47.70
TiO_2	0.79	0.75
Al_2O_3	12.20	11.60
Fe_2O_3 (tot)	9.28	8.99
MnO	0.17	0.16
MgO	9.86	9.93
CaO	7.40	9.49
Na_2O	2.92	2.43
K_2O	4.03	2.37
P_2O_5	0.65	0.68
LOI	4.08	4.54
Total	99.98	98.64
mg#	71	72
V	118	213
Cr	557	590
Ni	144	159
Rb	198	123
Sr	749	676
Y	23	28
Zr	275	244
Nb	10	12
Ba	765	1542
La	83	65
Ce	144	114
Hf	6.8	4.8

7.5.4 Petrology and Geochemistry of the Lamprophyres

The dykes are all shoshonitic types, including minettes, kersantites, and vogesites. In contrast to the lamprophyres of the Yilgarn Block, spessartites have not been recorded to date in the Superior Province.

Geochemically, the lamprophyres are characterized by primitive compositions, which are comparable to their Western Australian counterparts (see Sect. 7.4.5), with high mg# (> 70) and high concentrations of mantle-compatible elements (e.g. > 110 ppm V, ~ 550 ppm Cr, ~ 150 ppm Ni; Table 7.7). They are highly potassic (up to 5.6 wt % K_2O), and they are characterized by high concentrations of LILE (e.g.

Table 7.8.   Correlation matrix for precious metals (Au, Pd, Pt), Cu, and gold pathfinder elements (Sb, W) of lamprophyres from the Superior Province, Canada. All samples are derived from localities distal to gold deposits. Data from Wyman and Kerrich (1989b).

	Cu	Au	Pt	Pd	Sb	W
Cu	1					
Au	0.025	1				
Pt	0.878	-0.457	1			
Pd	0.819	-0.553	0.994	1		
Sb	-1.000	-0.052	-0.865	-0.803	1	
W	0.853	-0.500	0.999	0.998	-0.839	1

up to 200 ppm Rb, up to 750 ppm Sr, up to 1600 ppm Ba), and intermediate LREE (e.g. ~ 80 ppm La, ~ 140 ppm Ce), and low HFSE (e.g. ~ 0.8 wt % TiO_2, ~ 250 ppm Zr, ~ 10 ppm Nb, ~ 5 ppm Hf) concentrations (Table 7.7). Only a few published precious-metal contents of lamprophyres from the Superior Province are available (Wyman and Kerrich 1989a). They were derived from fresh, unaltered samples distal from mesothermal gold mineralization (Wyman and Kerrich 1989a) and no significant Au abundances were detected (Fig. 7.7). Table 7.8 shows the very low correlation between Au and the magmatic elements Cu, Pt, and Pd, and the normal pathfinder elements Sb and W although the database is limited. The significance of the Au contents of the rocks is discussed in Section 7.6, together with those of counterparts in the Yilgarn Block.

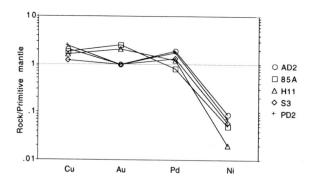

Fig. 7.7.   Abundances of chalcophile elements in lamprophyres from the Superior Province relative to the primitive mantle. All data are from lamprophyres distal to gold deposits. Normalizing factors after Brügmann et al. (1987). Data from Wyman and Kerrich (1989a).

7.6 Indirect Link Between Lamprophyres and Archaean Gold Mineralization

Many of the shoshonitic lamprophyres from the Superior Province, Canada, and the Yilgarn Block, Western Australia, plot in the postcollisional arc field (Fig. 6.15), consistent with their late-tectonic timing with respect to initial volcanism. However, discrimination is equivocal for several samples which plot in the continental arc field on the diagram. This could be explained by the high degree of alteration and extensive metamorphic recrystallization of most of the Archaean lamprophyres, which have been affected by a complex history of metamorphism, deformation, and hydrothermal alteration (e.g. Taylor et al. 1994), resulting in the mobilization of P and an increase in the Ce/P_2O_5 ratios. The P depletion is also displayed by pronounced negative anomalies in their spidergram patterns (Taylor et al. 1994).

The lamprophyric magmas were probably mantle-derived from between 50 and 120 km depth beneath the continental crust, whereas the gold mineralizing systems were probably confined to the continental crust (Wyman and Kerrich 1989a). Data from Wyman and Kerrich (1989a) and Taylor et al. (1994) indicate that the Archaean shoshonitic lamprophyres do not have intrinsically high Au concentrations. Sporadically anomalous Au abundances are restricted to samples from the vicinity of mesothermal gold deposits (Fig. 7.5b), and the Au values are decoupled from Cu and Pd peaks in primitive mantle-normalized distribution plots in the shoshonitic lamprophyres (Figs 7.5, 7.6), suggesting that anomalous Au contents are secondary, not primary, features. This is consistent with the poor correlation between Au and the magmatic elements Cu, Pt, and Pd (Tables 7.6, 7.8), although it is not supported by a good correlation between Au and the gold pathfinder elements As, Sb, and W.

Overall, the spatial but not genetic association of mesothermal gold deposits and shoshonitic magmatism since the late Archaean appears to be related to their common requirement of Phanerozoic-style subduction followed by oblique collision (e.g. Barley et al. 1989; Wyman 1990).

7.7 Synthesis of Indirect Associations

Although lamprophyres are spatially and temporally related to gold mineralization in several ancient terrains, the association is interpreted to be an indirect one.

Proterozoic mesothermal lode-gold mineralization and lamprophyre emplacement may be contemporaneous in several places, although no direct genetic relationships have been established yet. Both lamprophyres and mineralization commonly occur along major faults and shear zones which controlled their emplacement. The elevated Au contents in some Proterozoic lamprophyres from the vicinity of lode-gold deposits are decoupled from Cu and Pd peaks in primitive mantle-normalized distribution plots suggesting that the anomalous Au contents are secondary features. The

contamination en route to surface or caused by hydrothermal fluids overprinting the dykes after their emplacement.

Spatial associations also exist between Archaean mesothermal gold deposits and potassic lamprophyre dykes in both postcollisional arc and within-plate settings. Rock and Groves (1988a, 1988b) suggest that the volatile- and LILE-enriched fluids necessary to form a metasomatically enriched mantle capable of yielding potassic magmas would also be capable of transporting Au because they mimic hydrothermal fluids known to form gold deposits in the crust. However, recent studies (summarized in Sect. 8.3) argue against a direct genetic association, although lamprophyre intrusions may bracket the gold mineralization in time (Taylor et al. 1994; Yeats et al. 1999). The mesothermal gold deposits in these settings are not hosted by potassic igneous complexes, but they do occur along major faults and shear zones, which also represent the conduits for emplacement of lamprophyre intrusions (Wyman and Kerrich 1989a, 1989b). As for the Proterozoic lamprophyres, the elevated Au contents of some Archaean lamprophyres are decoupled from Cu and Pd peaks in primitive mantle-normalized distribution plots, suggesting that the anomalous Au contents are secondary features, caused by either crustal contamination by mineralized wallrocks or overprinting by later hydrothermal fluids.

Indirect associations between shoshonitic lamprophyres and mesothermal gold-antimony deposits have also been documented from the Permian Hillgrove district, New South Wales, Australia (Ashley et al. 1994). A similar direct relationship between shoshonitic lamprophyres and mesothermal copper-gold-bismuth mineralization has been recorded for the Tennant Creek area, although recent ^{40}Ar/^{39}Ar age dating indicates that the lamprophyres significantly postdate the mineralization (Duggan and Jaques 1996). Based on available geochemical and geochronological data, the widespread spatial association of gold deposits and lamprophyres, worldwide (e.g. Rock et al. 1988a, 1989), is probably an indirect, rather than genetic, association, related to the siting of both adjacent to crustal-scale deformation zones.

8 Halogen Contents of Mineralized Versus Unmineralized Potassic Igneous Rocks

8.1 Introduction

Previous studies have established the important role of halogens (Cl, F) for the transport of metals in ore deposits related to igneous rocks (Holland 1972; Kilinc and Burnham 1972; Gunow et al. 1980; Boudreau et al. 1986; Carten 1987; Webster and Holloway 1988, 1990; Richards et al. 1991; Webster 1992; Stanton 1994). It seems likely that the development of postmagmatic hydrothermal ore deposits depends less on the abundance of the ore metals than on the availability of appropriate mechanisms to concentrate, transport, and deposit the metals (Roegge et al. 1974). The presence of sufficient halogens such as Cl and/or F in the magmas appears to be the most important chemical parameter (Roegge et al. 1974). Chlorine largely controls the abundances of chlorophile ore and associated elements (e.g. Fe, Mn, Na, K, Cu) in saline aqueous fluids that exsolve from a magma (Webster 1992), and it also increases PGE solubilities in both sulphide and silicate melts (Peach et al. 1994). In the rock-forming minerals, Cl generally occupies the hydroxyl sites of micas, amphiboles, and apatites (Fuge et al. 1986). The strong affinity of the halogens for potassium, particularly in micas, can be explained in terms of their electronic configurations (Cocco et al. 1972).

The potential use of high Cl concentrations in mica phenocrysts as a prospecting tool to define mineralized volcanic and subvolcanic rocks was suggested by Stollery et al. (1971). Comparisons between the halogen contents of micas from mineralized and barren intrusions showed that micas from the former are generally characterized by higher Cl and F concentrations than those from the latter (Stollery et al. 1971; Kesler et al. 1975; Jiang et al. 1994; Loferski and Ayuso 1995). Although the preliminary studies did not focus on mica compositions of potassic igneous rocks, mica phenocrysts of unmineralized suites of these rocks are known to have higher halogen contents than those of K-poor intrusions (Kesler et al. 1975), in agreement with studies showing that the Cl solubility in volatile-rich melts increases with the alkalinity of the melt (Metrich and Rutherford 1992; Dixon et al. 1997). Recent studies have shown that alkaline magmas such as potassic melts are characteristically enriched in halogens (Bailey and Hampton 1990; Yang and Bodnar 1994; Foley 1994;

Table 8.1. Data sources of mica compositions from database MICA1. The number in square brackets refers to the number of analyses from that reference in the database. From Müller and Groves (1993).

1. Continental arcs	2. Postcollisional arcs	3b. Late oceanic arcs	4. Within-plate settings
American Cordillera (barren) Allan and Carmichael (1984) [12]	*Eastern Alps* D. Müller (unpubl. data, 1992) [6]	*Goonumbla, New South Wales* D. Müller et al. (1994) [5]	*Karinya Syncline, South Australia* Müller et al. (1993a) [7]
American Cordillera (mineralized) Kesler et al. (1975) [28]	*Grasberg, Indonesia* D. Müller (unpubl. data, 1993) [4]	*Ladolam, Papua New Guinea* D. Müller (unpubl. data, 1993) [9]	*Mount Bundey, Northern Territory, Australia* D. Müller (unpubl. data, 1992) [7]
Bingham, USA Parry et al. (1978) [7]	*Porgera, Papua New Guinea* D. Müller (unpubl. data, 1992) [3] Richards (1990b) [2]		
	Superior Province, Canada D. Müller (unpubl. data, 1992) [6]		
	Yilgarn Block, Western Australia D. Müller (unpubl. data, 1992) [19]		

Zhang et al. 1995), suggesting that the strongest Cl enrichments of magmatic-hydrothermal fluids, and ore metals complexed with Cl, occur in fluids exsolved from magmas that are relatively enriched in K_2O (Webster 1992). This is consistent with studies by Gammons and Williams-Jones (1997), which indicate that, at 500°C, gold solubility as $AuCl_2^-$ is highest for fluids that are oxidized ($SO_2/H_2S > 1$), highly saline, and potassium rich.

8.2   Erection of Database MICA1

The fact that most potassic igneous rocks contain mica phenocrysts, and that considerable mineral chemistry data have been published, led to compilation of a database of mica analyses — MICA1 — from mineralized and barren potassic igneous rocks to test their potential as exploration guides. The halogen contents of representative fresh mica phenocrysts from the investigated localities were both collated from the literature and directly analyzed using a ARL-SEMQ microprobe with attached WDS system at the Centre for Microscopy and Microanalysis, The University of Western Australia, Perth. All analyzed micas are homogeneous and unzoned. The available data are not filtered, and their sources are listed in Table 8.1. The significance of the data is discussed below.

8.3   Discussion

8.3.1 Behaviour of Halogens in Magmatic Hydrothermal Systems

During magma crystallization, trace elements and halogens partition between the melt and the crystallizing solids (Candela 1989; Cline and Bodnar 1991). At some stage during crystallization, bubbles of a magmatic aqueous phase nucleate and grow because the water concentration increases in the bulk melt as quartz and feldspar crystallize (Candela 1989; Lowenstern 1994). The exsolution of this aqueous fluid during the final stages of crystallization leads to stockwork veining due to hydraulic fracturing caused by expansion of the igneous pluton (Solomon and Groves 1994). This magmatic-hydrothermal transition occurs in response to decreasing pressure (first boiling) and crystallization (second boiling) as the melt approaches the surface (Cline and Bodnar 1991; Candela 1997). Chlorine has a pronounced preference for the aqueous fluid relative to the silicate melt and silicate minerals, implying that a Cl-bearing fluid would form during the crystallization of a H_2O-saturated, Cl-bearing magma (Kullerud 1995). Extreme enrichments in Cl and F may occur in these magmatic hydrothermal fluids during the end stages of crystallization (Webster and Holloway 1990). Additionally, significant quantities of ore elements may be partitioned into these hydrothermal fluid phases and be removed from the pluton (Candela 1989).

It is likely, therefore, that such Cl-rich hydrothermal fluids, which exsolve during magma crystallization, also transport gold and/or base metals (Kilinc and Burnham 1972), particularly since ore-metal solubility as chloride complexes in aqueous fluids is a strong function of their Cl content (Webster 1997). For example, the textures and geochemistry of plutons associated with Climax-type porphyry-molybdenum deposits indicate that Cl- and F-enriched magmatic hydrothermal fluids were primarily responsible for the transport of ore constituents (White et al. 1981; Webster and Holloway 1990). The very high salinities of the ore fluids in most porphyry copper-gold systems, as indicated by fluid inclusion studies (Roedder 1984), suggest that base metals were carried as chloride complexes, as was Au, at least at high temperatures (Hayba et al. 1985; Heald et al. 1987; Large et al. 1989).

Recent studies suggest that both Cu and Au preferentially partition into the volatile phase during magmatic devolatilization, being dissolved in that phase as chloride complexes (Hayashi and Ohmoto 1991). Cooling will eventually result in the disproportionation of the SO_2 to sulphate and sulphide species, thus leading to the precipitation of Cu-bearing sulphide minerals (Richards 1995). Although some Au may also be deposited with these sulphides, much of it will remain in solution by conversion from chloride to bisulphide complexing at lower temperatures (Hayashi and Ohmoto 1991). Most of the Au may be precipitated at shallower levels in the epithermal environment (Richards 1995), consistent with the decrease of Au grade downwards into an increasingly Cu-rich mineralization, as was originally proposed by Bonham and Giles (1983).

An important prerequisite for the hydrothermal extraction of Au from the magma is that removal of chalcophile elements from the melt by sulphide-liquid segregation should not occur before volatile saturation (Richards 1995). This condition may be achieved by the suppression of sulphur saturation through a high oxidation state of the magma (see Sect. 9.3.2), thus promoting the presence of sulphur as sulphate, not sulphide, within the melt (Richards 1995; Campbell et al. 1998).

Thus, Cl and F are important components of the hydrothermal system in that they represent an effective metal-transporting medium (Spear 1984; Loferski and Ayuso 1995). Hence, it may be no coincidence that many gold deposits tend to be associated with the more volatile-rich potassic and calc-alkaline magmas (Spooner 1993), where a magmatic connection is indicated (Richards 1995).

Chlorine is enriched, together with K_2O, in residual melts as a result of crystal-liquid differentiation of vapour-poor magmas, and a moderate correlation between these elements has been recorded for lavas from several localities (Anderson 1974). However, in vapour-rich lavas, Cl tends to partition into the vapour phase and hence is likely lost during degassing on magma ascent (A. Edgar, written comm. 1994). Fluorine contents increase regularly from tholeiites to potassic basalts (Aoki et al. 1981), and F is a significant element in potassic and ultrapotassic magmas (Edgar and Charbonneau 1991; Foley 1994). Experimental studies by Vukadinovic and Edgar (1993) suggest that F, in contrast to Cl (see Magenheim et al. 1995), behaves as a mantle-compatible element. Under mantle conditions, F tends to remain in the solid phases rather than partitioning into the first melt increments during partial melting (Vukadinovic and Edgar 1993). During crystallization, F is partitioned into the hydrous phenocrysts rather than remaining in the melt (Edgar et al. 1994). In accord with this, Kesler et al. (1975) and Naumov et al. (1998) have shown that the average whole-rock halogen contents of potassic intrusions (> 2 wt % K_2O) are higher (240 ppm Cl, 620 ppm F) than those for non-potassic (< 2 wt % K_2O) intrusions (160 ppm Cl, 380 ppm F). In addition, micas from mineralized intrusions tend to be more Cl-enriched than those from barren igneous rocks (Munoz 1984). The ions OH⁻, Cl⁻, F⁻, and K^+ are fixed preferentially in hydrous minerals, in particular phlogopites, under subsolidus conditions (Aoki et al. 1981; Spear 1984; Foley 1992). Phlogopite-bearing metasomatized mantle peridotites are considered to be source materials for primitive potassic magmas (Tatsumi and Koyaguchi 1989; Foley 1992), and phlogopites provide the major sites of Cl and F in potassic igneous rocks (Aoki et al. 1981; Edgar and Arima 1985; Edgar et al. 1994). Recent studies by Jiang et al. (1994) suggest that F is preferentially incorporated into micas with high Mg/Fe ratios, whereas Cl tends to be enriched in those with lower Mg/Fe ratios.

The ubiquitous presence of biotite in most porphyries indicates that water loss to hydrous phases and enhancement of the Cl/H_2O ratio occur in all productive systems (Cline and Bodnar 1991).

Water-leach analyses of biotites and phlogopites suggest that less than 10 wt % of Cl and 1 wt % of F are present in fluid inclusions. The remainder is apparently present in the mica structure (Kesler et al. 1975).

8.3.2 Halogen Contents of Mica in Potassic Igneous Rocks

The halogen contents of representative, fresh mica phenocrysts from potassic igneous rocks associated with mineralization from the investigated localities are plotted on a Cl versus F biaxial diagram in Figure 8.1. Representative data for micas from potassic igneous rocks interpreted to be genetically associated with epithermal gold mineralization are listed in Tables 8.2 (Ladolam, Papua New Guinea) and 8.3 (Porgera, Indonesia), and with porphyry copper-gold mineralization are listed in Tables 8.4 (Grasberg, Indonesia), and 8.5 (Goonumbla, Australia). Data from such rocks only spatially related to mineralization are listed in Tables 8.6 (Eastern Goldfields Province, Yilgarn Block, Australia) and 8.7 (Kirkland Lake gold district, Superior Province, Canada).

Mica phenocrysts from shoshonitic lamprophyres from the Eastern Goldfields Province (Kambalda area, Mount Morgans gold mine) and Kirkland Lake gold district are generally characterized by very low halogen (< 0.04 wt % Cl) concentrations (Fig. 8.1). This is consistent with the extremely low Cl concentrations (about 0.02 wt %) of apatite microphenocrysts in Superior Province lamprophyres (R. Kerrich, written comm. 1992). The low halogen contents of Yilgarn Block and Superior Province shoshonitic lamprophyres, when compared to those from mineralized high-K igneous rocks from equivalent postcollisional arc settings (Fig. 8.1), lend further credence to the arguments that the lamprophyres are not genetically related to the mesothermal gold mineralization as originally suggested by Rock et al. (1987) and Rock and Groves (1988a), but have an indirect association (Wyman and Kerrich 1989b; Taylor et al. 1994; see Sect. 7.6).

In contrast, mica phenocrysts from potassic igneous rocks which host epithermal gold or porphyry copper-gold mineralization are characterized by elevated halogen abundances (> 0.04 wt % Cl). Examples are micas from the high-K igneous rocks associated with the epithermal gold deposits at Ladolam (up to 0.29 wt % Cl) and Porgera (up to 0.09 wt % Cl), and from the Grasberg (up to 0.24 wt %) and Goonumbla (up to 0.14 wt % Cl) porphyry copper-gold deposits.

Where mineralization is interpreted to be genetically associated with potassic igneous rocks, the host rocks were generated in continental, postcollisional or late oceanic arcs (cf. Müller and Groves 1993). These three settings are the only tectonic settings in which mineralization genetically related to potassic igneous rocks is currently known to occur. The high Cl concentrations (> 0.04 wt % Cl) in mica phenocrysts from all mineralized potassic igneous rocks further support a direct genetic relationship between magmatism and mineralizing fluids in these three specific tectonic settings.

Micas from the shoshonitic host rocks of the Ladolam epithermal gold (Table 8.2), and the Goonumbla porphyry copper-gold deposit (Table 8.5), both interpreted to have been generated in a late oceanic-arc setting (Müller et al. 1994), are enriched in Cl *and* F (Fig. 8.1). This might be a specific feature of mineralized potassic igneous rocks from this setting, but it awaits confirmation from data on deposits such as the Emperor epithermal gold deposit, Fiji, which is in a similar tectonic setting.

Importantly, analyzed unmineralized potassic igneous suites from continental,

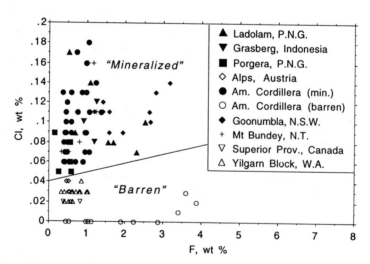

Fig. 8.1.   Chlorine and F compositions of mica phenocryst from potassic igneous rocks from barren and mineralized environments (data sources are listed in Table 8.1). Adapted from Müller and Groves (1993).

postcollisional, and late oceanic arcs contain low-Cl (±F) mica phenocrysts, and hence contrast with those from mineralized settings (Fig. 8.1). It should be noted, however, that analyzed phenocrysts from potassic igneous rocks from within-plate settings (e.g. Mount Bundey suite, Fig. 8.1) have high Cl contents equivalent to those of mineralized high-K rocks from continental, postcollisional, and late oceanic arc settings, although the rocks are interpreted to have no genetic relationship to gold mineralization in the area. It may also be significant that some potassic igneous rocks from within-plate settings (e.g. Karinya Syncline, Sect. 5.3) have intrinsically high Au contents although no direct relationships to mineralization have yet been established.

Potassic igneous rocks from the Mariana Arc (Sect. 4.3), an initial oceanic arc, do not contain mica phenocrysts (Dixon and Batiza 1979) and cannot, therefore, be assessed in terms of halogen contents.

Mica phenocrysts from potassic igneous rocks generated in within-plate tectonic settings can contain elevated halogen concentrations. Studies by Naumov et al. (1998) suggest also that potassic rocks with high TiO_2 contents (~ 3.23 wt %), which are typical for those from within-plate tectonic settings, have primary enrichments in F compared to the average F concentrations of basic melts. However, within-plate tectonic settings may have no known mineralization or there may be only an indirect association with known mineralization (e.g. Mount Bundey igneous suite and Tom's Gully gold deposit).

8.3.3 Significance of Halogen Data

From the above discussion, it appears that the halogen contents of mica phenocrysts from high-K igneous suites can be used, with due caution, as a measure of gold-copper mineralization potential in continental, postcollisional, and late oceanic arcs (Müller and Groves 1993). The threshold value for Cl concentration for such micas is about 0.04 wt % or 400 ppm, but F is a poor discriminant, being enriched up to about 4.0 wt % in potassic rocks from both mineralized and unmineralized environments (Fig. 8.1).

Table 8.2. Microprobe (WDS) analyses of mica phenocrysts from shoshonitic rocks from the Ladolam gold deposit, Lihir Island, Papua New Guinea. Ox. form. = oxygen formula. Samples were provided by H. Hoogvliet, Kennecott Exploration. Sample numbers refer to specimens held in the Museum of the Department of Geology and Geophysics, The University of Western Australia.

Sample no.:	119429	119429	119430	119430
wt %				
SiO_2	37.93	37.39	38.12	37.18
TiO_2	2.86	1.16	4.30	4.38
Al_2O_3	12.74	15.01	14.32	14.05
FeO (tot)	12.52	10.93	11.17	11.45
Cr_2O_3	-	-	-	-
MnO	-	-	-	-
MgO	16.63	16.62	16.43	16.74
NiO	-	-	-	-
BaO	0.20	0.25	0.23	0.21
CaO	0.04	0.03	0.08	0.08
SrO	0.48	0.14	0.15	0.15
K_2O	9.35	9.76	9.22	9.06
Na_2O	0.38	0.41	0.50	0.34
Cl	0.08	0.07	0.17	0.29
F	1.89	2.41	0.70	1.21
Total	94.31	93.18	95.10	94.63
mg#	75	78	77	77
Ox. form.	22	22	22	22
Atoms				
Si	5.770	5.805	5.653	5.580
Ti	0.327	0.135	0.479	0.494
Al	2.284	2.747	2.502	2.484
Fe	1.593	1.420	1.384	1.437
Cr	-	-	-	-
Mn	-	-	-	-
Mg	3.767	3.845	3.628	3.741
Ni	-	-	-	-
Ba	0.008	0.003	0.009	0.011
Ca	0.007	0.005	0.013	0.013
Sr	0.042	0.006	0.003	0.010
K	1.814	1.932	1.744	1.734
Na	0.113	0.123	0.144	0.100
Cl	-	-	-	-
F	-	-	-	-
Total	15.725	16.021	15.559	15.604

119430	119431	119431	119431
36.53	37.22	37.00	36.59
4.48	4.16	4.44	4.20
13.99	13.22	13.91	13.66
11.06	12.86	12.62	12.74
-	-	-	-
-	-	-	-
16.50	15.67	15.63	15.78
-	-	-	-
0.23	0.23	0.23	0.29
0.04	0.03	0.03	0.03
0.14	0.13	0.26	0.14
9.13	9.42	9.39	9.43
0.38	0.27	0.39	0.26
0.10	0.11	0.09	0.14
2.62	0.63	0.61	1.21
94.10	93.69	94.35	93.95
78	74	74	74
22	22	22	22
5.632	5.744	5.663	5.659
0.519	0.483	0.511	0.489
2.543	2.405	2.510	2.489
1.426	1.659	1.615	1.647
-	-	-	-
-	-	-	-
3.790	3.602	3.563	3.636
-	-	-	-
-	0.004	0.014	0.018
0.006	0.005	0.004	0.001
0.011	0.008	0.023	0.004
1.796	1.854	1.834	1.860
0.115	0.081	0.114	0.078
-	-	-	-
-	-	-	-
15.838	15.845	15.851	15.881

Table 8.3.   Microprobe (WDS) analyses of mica phenocrysts from potassic igneous rocks from the Porgera gold deposit, Papua New Guinea. Ox. form. = oxygen formula. Samples are from the private collection of N.M.S. Rock, and other data are from Richards (1990b).

Sample no.:	PO13[a]	PO14[a]	PO14[a]	RJR-46-A[b]	RJR-21-A[b]
wt %					
SiO_2	35.62	36.07	37.25	36.13	36.63
TiO_2	3.78	3.97	3.56	3.77	2.70
Al_2O_3	14.58	14.94	14.94	14.31	15.23
FeO (tot)	15.70	14.64	13.19	14.41	11.50
Cr_2O_3	0.03	0.03	0.03	0.03	0.09
MnO	0.39	0.27	0.24	0.27	0.13
MgO	14.84	15.86	17.07	16.93	18.64
NiO	0.09	0.10	0.09	-	-
BaO	0.18	0.18	0.18	0.20	0.80
CaO	0.03	0.04	0.03	0.02	0.01
SrO	0.16	0.16	0.15	-	-
K_2O	8.80	9.12	8.64	8.76	8.89
Na_2O	0.89	0.91	1.12	1.08	0.76
Cl	0.06	0.06	0.08	0.05	0.09
F	0.59	0.55	0.55	0.26	0.14
Total	95.48	96.66	96.89	96.21	95.63
mg#	68	71	75	73	79
Ox. form.	22	22	22	22	22
Atoms					
Si	5.410	5.380	5.475	5.389	5.423
Ti	0.432	0.445	0.393	0.423	0.300
Al	2.609	2.626	2.587	2.515	2.658
Fe	1.994	1.826	1.620	1.797	1.424
Cr	-	0.002	0.003	0.004	0.010
Mn	0.050	0.033	0.029	0.034	0.017
Mg	3.359	3.524	3.737	3.763	4.114
Ni	0.008	0.011	0.011	-	-
Ba	0.005	0.006	0.006	0.012	0.046
Ca	0.002	0.007	0.004	0.003	-
Sr	-	-	-	-	-
K	1.704	1.734	1.619	1.667	1.678
Na	0.261	0.262	0.319	0.313	0.217
Cl	-	-	-	-	-
F	-	-	-	-	-
Total	15.834	15.856	15.803	16.052	15.979

[a] Private collection of N.M.S. Rock.
[b] From Richards (1990b).

Table 8.4.   Microprobe (WDS) analyses of mica phenocrysts from potassic igneous rocks from the Grasberg copper-gold deposit, Indonesia. Ox. form. = oxygen formula.

Sample no.:	DM1	DM1	DM1	DM1
wt %				
SiO_2	35.56	35.24	35.37	35.44
TiO_2	5.01	5.02	5.27	4.61
Al_2O_3	13.59	13.08	13.58	14.07
FeO (tot)	16.96	17.32	17.13	16.97
Cr_2O_3	-	-	-	-
MnO	-	-	-	-
MgO	12.65	12.95	13.56	13.67
NiO	-	-	-	-
BaO	0.21	0.32	0.23	0.20
CaO	0.08	0.03	0.03	0.03
SrO	0.09	0.11	0.11	0.09
K_2O	9.43	9.46	9.67	9.79
Na_2O	0.40	0.33	0.26	0.28
Cl	0.12	0.26	0.08	0.10
F	1.30	1.42	1.23	1.31
Total	95.40	95.54	96.52	96.56
mg#	63	63	65	65
Ox. form.	22	22	22	22
Atoms				
Si	5.498	5.483	5.416	5.418
Ti	0.582	0.587	0.606	0.529
Al	2.475	2.397	2.449	2.535
Fe	2.192	2.253	2.192	2.168
Cr	-	-	-	-
Mn	-	-	-	-
Mg	2.913	3.001	3.093	3.114
Ni	-	-	-	-
Ba	0.013	0.019	0.014	0.012
Ca	0.013	0.001	0.001	0.005
Sr	0.006	-	-	0.005
K	1.859	1.877	1.888	1.909
Na	0.120	0.099	0.076	0.083
Cl	-	-	-	-
F	-	-	-	-
Total	15.671	15.717	15.735	15.778

Table 8.5.   Microprobe (WDS) analyses of mica phenocrysts from shoshonitic rocks from the Goonumbla copper-gold deposit, New South Wales, Australia. Ox. form. = oxygen formula. Sample numbers refer to specimens held in the Museum of the Department of Geology and Geophysics, The University of Western Australia.

Sample no.:	119086	119086	119086	119100	119102
Origin:	E26N	E26N	E26N	E26N	E26N
Rock type:	Monzonite	Monzonite	Monzonite	Trachyte	Trachyte
wt %					
SiO_2	36.81	36.10	37.12	36.53	35.46
TiO_2	4.36	4.22	4.00	2.83	3.45
Al_2O_3	14.07	14.35	14.17	16.42	16.59
Cr_2O_3	0.02	0.02	0.02	0.02	0.02
FeO (tot)	11.27	11.65	11.50	13.03	13.33
MnO	0.11	0.10	0.16	0.16	0.22
MgO	17.31	16.57	16.35	15.65	15.32
CaO	0.03	0.03	0.03	0.03	0.03
Na_2O	0.43	0.42	0.44	0.29	0.29
K_2O	9.15	9.20	9.10	8.85	9.47
SrO	0.14	0.15	0.16	0.15	0.16
BaO	0.18	0.28	0.19	0.14	0.14
Cl	0.10	0.10	0.14	0.11	0.12
F	2.67	3.90	3.18	1.58	1.42
Total	95.85	95.88	95.83	95.85	95.53
mg#	78	77	77	74	73
Ox. form.	22	22	22	22	22
Atoms					
Si	5.244	5.136	5.290	5.242	5.120
Ti	0.468	0.452	0.428	0.306	0.374
Al	2.362	2.406	2.380	2.778	2.824
Cr	0.002	0.002	0.002	0.002	0.002
Fe	1.208	1.248	1.234	1.406	1.448
Mn	0.014	0.012	0.020	0.020	0.026
Mg	3.674	3.514	3.472	3.346	3.296
Ca	0.004	0.004	0.004	0.004	0.004
Na	0.118	0.116	0.122	0.080	0.082
K	1.664	1.670	1.654	1.648	1.744
Sr	0.012	0.012	0.014	0.012	0.014
Ba	0.010	0.016	0.010	0.008	0.008
Cl	-	-	-	-	-
F	-	-	-	-	-
Total	14.796	14.604	14.646	14.868	14.958

Table 8.6.   Microprobe (WDS) analyses of mica phenocrysts from shoshonitic lamprophyres from the Kambalda gold province and from Mount Morgans gold deposit, Yilgarn Block, Western Australia. Ox. form. = oxygen formula. Samples from Mount Morgans were provided by R.M. Vielreicher, The University of Western Australia, and other samples are held in the Museum of the Department of Geology and Geophysics, The University of Western Australia.

Sample no.:	108370	108512	RV5 [a]	RV5 [b]	B64	B10
Locality	Kambalda area	Kambalda area	Mount Morgans	Mount Morgans	Mount Morgans	Mount Morgans
wt %						
SiO_2	37.07	37.93	40.66	40.09	36.54	38.12
TiO_2	2.27	2.31	1.37	1.84	2.28	1.89
Al_2O_3	17.98	14.29	12.77	12.96	15.27	13.87
FeO (tot)	11.08	15.79	2.88	10.18	16.76	15.43
Cr_2O_3	0.03	0.10	0.47	0.45	0.13	0.10
MnO	0.16	0.08	0.05	0.09	0.17	0.20
MgO	14.94	14.84	24.54	18.96	13.19	15.15
NiO	0.10	0.06	0.70	0.55	0.10	0.11
BaO	0.22	0.37	0.17	0.19	0.19	0.22
CaO	0.03	0.03	0.03	0.03	0.07	0.03
SrO	0.13	0.13	0.14	0.14	0.15	0.15
K_2O	10.14	9.35	10.68	10.27	9.06	9.81
Na_2O	0.10	0.21	0.03	0.03	0.09	0.04
Cl^-	0.03	0.03	0.03	0.03	0.02	0.03
F	1.02	0.41	0.81	0.64	0.58	0.62
Total	94.87	95.76	94.99	96.18	94.36	95.51
mg#	75	68	95	81	64	69
Ox. form.	22	22	22	22	22	22
Atoms						
Si	5.518	5.695	5.847	5.868	5.601	5.752
Ti	0.254	0.260	0.147	0.203	0.262	0.215
Al	3.153	2.528	2.163	2.235	2.758	2.465
Fe	1.378	1.981	0.347	1.246	2.148	1.946
Cr	0.001	0.012	0.053	0.052	0.016	0.011
Mn	0.019	0.010	0.006	0.011	0.022	0.025
Mg	3.313	3.319	5.256	4.134	3.012	3.406
Ni	0.012	0.007	0.081	0.065	0.012	0.006
Ba	0.011	0.022	0.006	0.004	0.007	0.013
Ca	0.002	0.030	-	0.001	0.012	0.004
Sr	0.011	0.011	0.007	0.007	0.002	-
K	1.924	1.790	1.958	1.916	1.772	1.887
Na	0.028	0.060	0.004	-	0.027	0.012
Cl	-	-	-	-	-	-
F	-	-	-	-	-	-
Total	15.624	15.725	15.875	15.742	15.651	15.742

[a] Core.   [b] Rim.

Table 8.7.   Microprobe (WDS) analyses of mica phenocrysts from shoshonitic lamprophyres from the Kirkland Lake gold district, Superior Province, Canada. Ox. form. = oxygen formula. Samples are from the private collection of N.M.S. Rock.

Analysis no.:	1	2	3	4	5	6
wt %						
SiO_2	38.89	38.35	37.47	36.46	35.99	35.77
TiO_2	1.76	2.01	1.73	3.39	3.16	3.13
Al_2O_3	15.05	15.05	15.33	13.92	14.14	14.07
FeO (tot)	12.78	13.04	12.99	16.68	16.70	16.55
Cr_2O_3	0.09	0.17	0.13	0.02	0.04	0.03
MnO	0.18	0.11	0.16	0.18	0.24	0.24
MgO	16.49	15.89	15.76	14.24	14.24	14.32
NiO	0.10	0.09	0.07	0.06	0.06	0.06
BaO	0.35	0.27	0.35	0.52	0.53	0.68
CaO	0.04	0.12	0.03	0.03	0.03	0.03
SrO	0.16	0.10	0.13	0.11	0.10	0.10
K_2O	10.03	9.52	10.01	9.27	9.44	9.32
Na_2O	0.04	0.09	0.09	0.22	0.17	0.18
Cl	0.02	0.04	0.04	0.04	0.05	0.04
F	0.38	0.36	0.33	0.37	0.44	0.34
Total	96.31	95.10	94.24	94.91	95.14	94.43
Ox. form.	22	22	22	22	22	22
Atoms						
Si	5.731	5.712	5.646	5.557	5.514	5.503
Ti	0.195	0.225	0.195	0.388	0.363	0.362
Al	2.612	2.641	2.722	2.499	2.553	2.551
Fe	1.574	1.623	1.637	2.125	2.138	2.128
Cr	0.010	0.019	0.015	0.002	0.004	0.003
Mn	0.022	0.014	0.020	0.023	0.031	0.031
Mg	3.618	3.526	3.537	3.232	3.249	3.281
Ni	0.011	0.010	0.008	0.004	0.004	0.004
Ba	0.020	0.016	0.021	0.031	0.031	0.041
Ca	0.006	0.019	0.002	0.004	0.003	0.001
Sr	0.013	0.008	0.011	0.003	0.004	0.009
K	1.884	1.808	1.923	1.802	1.845	1.827
Na	0.010	0.025	0.026	0.064	0.050	0.054
Cl	-	-	-	-	-	-
F	-	-	-	-	-	-
Total	15.710	15.649	15.764	15.738	15.792	15.799

9 Implications for Mineral Exploration

9.1 Introduction

Some of the observations discussed in previous chapters can be incorporated into the development of a strategy for the exploration for epithermal gold and porphyry copper-gold deposits, since many of these deposits are spatially associated with, or hosted by, potassic igneous rocks. The points listed below may be particularly useful.

9.2 Area Selection

9.2.1 Composition of Host Rocks

Rocks of the alkaline suite are generally more prospective for giant and bonanza gold deposits than normal calc-alkaline andesites (Sillitoe 1993). Specifically, four of the nine largest epithermal gold-silver deposits and four of the ten largest porphyry copper-gold deposits are associated with potassic igneous rocks (J. Hedenquist, pers. comm. 1996; Sillitoe 1997). The high-K igneous rocks only comprise between 5 and 10 % of arc rocks, yet are associated with 40 % of the largest epithermal and porphyry deposits, clearly indicating their selective importance.

Epithermal gold and porphyry copper-gold deposits are abundant in convergent-plate margin settings, reflecting their direct genetic association with high-K calc-alkaline and shoshonitic magmatism (Clark 1993). The occurrence of potassic igneous rocks clearly indicates a potentially high prospectivity for epithermal gold and porphyry copper-gold deposits in that area (Müller and Groves 1993). These rocks are defined by high K_2O (> 1 wt % at 50 wt % SiO_2), high LILE and low HFSE (see Chaps 2, 3) contents. Importantly, from a field exploration viewpoint, these rocks can commonly be recognized in hand specimen by their porphyritic texture with abundant phenocrysts of biotite, phlogopite, and/or amphibole. In addition, due to

to their high volatile contents, ore-related potassic igneous rocks may be characterized by the occurrence of interconnected miarolitic cavities (Candela 1997). The
hypabyssal lamprophyre intrusions are typified by a lack of plagioclase phenocrysts,
reflecting the relatively high water content of the magmas, which suppressed the
crystallization of feldspar (Carmichael et al. 1996). However, extrusive high-K rocks
can have plagioclase phenocrysts. Potassic igneous rocks are also characterized mi-
croscopically by apatite microphenocrysts and a groundmass that is dominated by
potassic feldspar or leucite.

9.2.2 Tectonic Setting

The definition of the tectonic setting in which the high-K host rocks have been generated is essential, since known epithermal gold and/or porphyry copper-gold mineralization appears to be restricted to the three subduction-related tectonic settings:
continental, postcollisional, and late oceanic arcs.

   The potassic igneous rocks themselves can be used to help discriminate the tectonic setting, as discussed in Chapter 3, via the use of hierarchical chemical discrimination diagrams (Figs 3.9, 3.10).

9.3   Prospect Evaluation

9.3.1 Favourable Tectonic Elements on the Prospect Scale

There is considerable evidence that many epithermal gold and porphyry copper-
gold deposits which have been generated in late oceanic arcs are located either within,
or at the margin of, collapsed caldera structures. Examples are the Ladolam gold
deposit, the Emperor gold deposit, and the Goonumbla copper-gold deposit (Müller
and Groves 1993). In addition, the intersections of major lineaments may be favourable tectonic elements on the prospect scale, as, for example, at Emperor (Setterfield
1991). Other major structures which favour mineralization (Sillitoe 1993) are high-
angle reverse faults (e.g. El Indio), strike-slip faults (e.g. epithermal gold deposit at
Baguio, Philippines), and reactivated collisional sutures (e.g. Porgera). The mineralization in Upper Eocene to Lower Oligocene porphyry-copper deposits in northern Chile (e.g. Chuquicamata, El Salvador, La Escondida, Zaldivar) appears to be
controlled by a major dextral transpressive structure (cf. Clark 1993). Thus, both
synvolcanic structures and post- to late-volcanic structures may control mineralization in prospective environments.

9.3.2 High Oxidation State of the Magmas

The oxidation state of the lithospheric upper mantle is heterogeneous, with a variation in the oxygen fugacity (fO_2) of at least four log units (Ballhaus et al. 1990; Ballhaus 1993).

Lithosphere above subduction zones is commonly assumed to be more oxidized than other mantle regimes (W.R. Taylor, pers. comm. 1992) because of its infiltration by slab-derived fluids generated from dehydration and decarbonation reactions (Arculus 1985; Haggerty 1990; Lange and Carmichael 1990). Phlogopite phenocrysts from subduction-related potassic igneous rocks have the highest Fe^{3+}/Fe^T ratios (0.77–0.87), which is consistent with the high magmatic fO_2 inferred for these magmas (Feldstein et al. 1996). The dissociation of H_2O and the release of H should enrich the system in oxygen at an early stage before the onset of partial melting (Abdel-Rahman 1994). However, recent experimental studies by Moore et al. (1995) indicate that dissolved water does not measurably affect the redox state of iron in natural melts. Therefore, the high oxygen fugacities which are commonly recorded in water-rich magmas (e.g. Lange and Carmichael 1990) are probably a record of other processes that have imposed a high fO_2 upon the melt, and are not a reflection of the amount of dissolved water (Moore et al. 1995). For example, studies by Dixon et al. (1997) show that the lowest degree partial melts have the highest relative magmatic fO_2 and this decreases with increasing extent of melting. The most oxidized basaltic melts are potassic lamprophyres such as minettes from mature continental arcs (Ballhaus 1993) and primitive basalts from oceanic arcs (Ballhaus 1993). There is also a positive correlation between the average volatile content in basalts and their calculated fO_2. Thus, those basalts from oceanic arcs which are most oxidized also have the highest abundances of volatile phases such as magmatic water (Ballhaus 1993).

At magmatic temperatures, fO_2 is a major factor controlling element partitioning (e.g. Ballhaus 1995; Blevin and Chappell 1996). At high values of fO_2, iron is partially transformed into Fe^{3+} and oxide minerals such as magnetite crystallize in preference to Fe^{2+}-bearing silicate minerals (Haggerty 1990). The high alkali content of potassic melts increases their Fe^{3+}/Fe^{2+} ratios, thus resulting in a higher oxygen fugacity (Wyborn 1994). A high fO_2 of the magmas, as indicated by significant concentrations of magnetite (> 5 vol. %) in their crystallized products, favours precipitation of large quantities of gold (Sillitoe 1979). So it is no coincidence that the potassic host rocks of many gold-rich porphyry copper deposits are characterized by appreciable magnetite. Examples are the high-K rocks hosting the porphyry copper-gold deposits at Bajo de la Alumbrera, Argentina (Stults 1985), Goonumbla, New South Wales, Australia (Müller et al. 1994), Grasberg, Indonesia (I. Kavalieris, pers. comm. 1996), and Marian, Philippines (Sillitoe 1979), which all contain up to 5 vol. % magnetite. These magnetite-rich igneous rocks result in a high magnetic susceptibility, generating magnetic responses of up to 4500 gammas, and may, therefore, be identified in airborne magnetic surveys. However, in areas where the rocks are strongly affected by alteration, the effects of supergene martitization (i.e. alteration of magnetite to haematite) can minimize this effect in the vicinity of mineralization.

At constant temperatures, as fO_2 increases, the concentration of dissolved sulphide (S^{2-}) decreases, whereas the dissolved sulphate (SO_4^{2-}) increases (Carmichael and Ghiorso 1986; Campbell et al. 1998), thus minimizing the loss of precious metals into sulphide phases (Richards 1995; Sect. 5.2). In agreement with this, MacInnes (1996) and Kesler (1997) have pointed out that alkaline magmas also can dissolve more sulphate, thus making it unlikely for the magma to become saturated with metal sulphides. Alkali-rich, basic magmas, particularly potassic melts, may contain sulphate minerals (e.g. anhydrite) and are, therefore, apparently considerably more oxidized than, for example, tholeiites (Haggerty 1990). The abundance of late anhydrite veins in the host rocks at the Ladolam (Sect. 6.3.1), Grasberg (Sect. 6.5.1), and Porgera (J. Standing, pers. comm. 1993; Sect. 6.5.3) gold deposits may be consistent with a high fO_2 for these potassic intrusions, and the occurrence of anhydrite may be a further guide to mineralization at the prospect scale.

9.3.3 Elevated Halogen Contents of the Magmas

The important role of halogens, such as Cl and F, for the transport of metals in ore deposits related to igneous intrusions is discussed in Chapter 8. Experimental studies of Cl partitioning in granitic systems suggest that the strongest enrichment of magmatic-hydrothermal fluids in Cl and ore metals complexed with Cl will most likely be in fluids exsolved from magmas that are relatively enriched in K_2O (Webster 1992). The halogen contents of mica phenocrysts from high-K igneous suites may be used as a measure of gold-copper mineralization potential in continental, postcollisional, and late oceanic arcs. Those potassic igneous rocks whose mica phenocrysts have Cl contents > 0.04 wt % and F contents > 0.5 wt % can be regarded as highly prospective. However, caution must be used in universal application of the halogen contents of high-K rocks as a tool in mineral exploration because such rocks analysed to date from within-plate settings also have high halogen contents, although no direct genetic relationships to gold-copper deposits have yet been established (Müller and Groves 1993).

10   Characteristics of Some Gold-Copper Deposits Associated with Potassic Igneous Rocks

The tables in this chapter summarize the characteristic features of some gold-copper deposits associated with potassic igneous rocks for which substantial information is available.

10.1   Abbreviations

Abbreviations for ore minerals: Apy, arsenopyrite; Bn, bornite; Cal, calaverite; Cc, chalcocite; Cinn, cinnabar; Cov, covellite; Cpy, chalcopyrite; Dig, digenite; En, enargite; Gn, galena; Mar, marcasite; Mol, molybdenite; Po, pyrrhotite; Py, pyrite; Sch, scheelite; Sl, sphalerite; Stib, stibnite; Ten, tennantite; Tet, tetrahedrite.

Abbreviations for silicate minerals: Af, alkali feldspar; An, analcite; Ap, apatite; Bio, biotite; Cpx, clinopyroxene; Foids, felspathoid minerals; Hb, hornblende; Mt, magnetite; Ol, olivine; Opx, orthopyroxene; Phl, phlogopite; Pl, plagioclase; Qz, quartz; Ti, titanite.

10.2   Tables of Deposit Characteristics

The deposit tables are arranged alphabetically for ease of use but the following lists indicate their tectonic settings.

Continental arcs: Andacollo, Bajo de la Alumbrera, Bingham, Choquelimpie, Cripple Creek (transitional), El Indio, Maricunga Belt, Summitville, Twin Buttes.
Postcollisional arcs: Grasberg (?), Kirkland Lake Mining District, Misima, Mount Kare, Mount Morgans, Ok Tedi, Porgera, Wiluna.
Late oceanic arcs: Cadia, Dinkidi, Emperor, Goonumbla, Ladolam, Woodlark Island.
Within-plate settings: Cripple Creek (transitional), Tom's Gully.

10.2.1   Andacollo, Chile

Deposit	Andacollo
Mineralization	Porphyry copper-gold, highest grades associated with potassic alteration
Estimated reserves	About 300 million tonnes at 0.7 wt % Cu and 0.25 ppm Au (about 2.1 million tonnes copper and 75 tonnes gold)
Location	Central Chile; 400 km north of Santiago
Host rocks	Andesite	Dacite
Texture	Porphyritic	Porphyritic
Structure	Fine to medium grained	Medium grained
Mineralogy	Phenocrysts: Pl-Hb-Bio±Af	Phenocrysts: Qz-Pl-Hb-Bio
	Groundmass: Pl-Af-Qz	Groundmass: Qz-Af-Pl
Geochemistry	Potassic calc-alkaline to shoshonitic, high LILE, moderate LREE, low HFSE	
Ore minerals	Py±Cpy±Sl±Gn±Cinn
Alteration	Potassic and propylitic
Age	Host rocks: Cretaceous (104 ± 3 to 98 ± 2 Ma)
	Mineralization: coeval with host rocks
Tectonic setting	Continental arc
Structural features	-
References	Reyes (1991)

10.2.2   Bajo de la Alumbrera, Catamarca Province, Argentina

Deposit	Bajo de la Alumbrera
Mineralization	Porphyry copper-gold
Estimated reserves	806 million tonnes at 0.53 wt % Cu and 0.64 ppm Au
Location	Catamarca Province, Northwest Argentina; 100 km northeast of Belen and 300 km southwest of Tucuman
Host rocks	Dacite
Texture	Porphyritic
Structure	Fine to medium grained
Mineralogy	Phenocrysts: Pl-Qz-Bio±Hb±Ap
	Groundmass: Af-Pl-Qz
Geochemistry	Potassic calc-alkaline to shoshonitic, high LILE, moderate LREE, low HFSE
Ore minerals	Py-Cpy±Bn
Alteration	Potassic, phyllic, and propylitic
Age	Host rocks: 7.9 Ma
	Mineralization: coeval with host rocks
Tectonic setting	Continental arc
Structural features	Mineralization is located within a tectonic depression and crosscut by late northwest-striking andesite dykes
References	Stults (1985), H. Salgado (pers. comm. 1994), Müller and Forrestal (1998), D. Keough (written comm. 1998)

10.2.3   Bingham, Utah, USA

Deposit	Bingham
Mineralization	Porphyry copper-molybdenum±gold±silver, highest grades associated with potassic alteration
Production	About 1300 million tonnes at 0.85 wt % Cu (about 11 million tonnes copper)
Location	Bingham Canyon, Utah, USA
Host rocks	Monzonite
Texture	Equigranular
Structure	Fine grained
Mineralogy	Af-Pl-Qz-Cpx-Bio±Ap±Mt
Geochemistry	Highly potassic
Ore minerals	Py-Cpy; as stockwork veining or disseminated
Alteration	Propylitic and potassic
Age	Host rocks: Eocene (39.8 ± 0.4 to 38.8 ± 0.4 Ma)
	Mineralization: coeval with host rocks
Tectonic setting	Continental arc
Structural features	Mineralization and igneous intrusions are controlled by the intersection of northeast-trending faults with the Copperton anticline
References	Lanier et al. (1978a, 1978b), Warnaars et al. (1978), Einaudi (1982)

10.2.4   Cadia, New South Wales, Australia

Deposit	Cadia	
Mineralization	Porphyry copper-gold	
Estimated reserves	About 260 tonnes gold and 1.1 million tonnes copper	
Location	New South Wales, Australia; 25 km south of Orange	
Host rocks	Trachyandesite	Quartz monzonite
Texture	Porphyritic	Porphyritic
Structure	Medium grained	Medium grained
Mineralogy	Phenocrysts: Pl-Hb	Phenocrysts: Pl-Qz-Bio
	Groundmass: Af-Pl-Hb	Groundmass: Af-Pl
Geochemistry	Shoshonitic, with high LILE, moderate LREE, low HFSE	
Ore minerals	Py-Cpy- native gold	
Alteration	Potassic, phyllic, and propylitic	
Age	Host rocks: Upper Ordovician	
	Mineralization: coeval with host rocks	
Tectonic setting	Late oceanic arc	
Structural features	Deposit is crosscut by northeast-striking normal faults	
References	Welsh (1975), J. Holliday (pers. comm., 1995), F. Leckie (pers. comm., 1995)	

10.2.5   Choquelimpie, Chile

Deposit	Choquelimpie
Mineralization	Epithermal gold-silver
Estimated reserves	About 6.7 million tonnes at 2.23 ppm Au and 87 ppm Ag (about 15 tonnes gold and 47 tonnes silver)
Location	Northern Chile; about 115 km east of Arica
Host rocks	Andesite	Dacite
Texture	Porphyritic	Porphyritic
Structure	Fine to medium grained	Medium grained
Mineralogy	Phenocrysts: Pl-Hb-Bio±Af	Phenocrysts: Qz-Pl-Hb-Bio
	Groundmass: Pl-Af-Qz	Groundmass: Pl-Af-Qz
Geochemistry	Potassic calc-alkaline, high LILE, moderate LREE, low HFSE	
Ore minerals	Py-Sl-Gn-Tet±Po± native silver	
Alteration	Argillic and siliceous	
Age	Host rocks: Miocene	
	Mineralization: coeval with host rocks	
Tectonic setting	Continental arc	
Structural features	Magmatism controlled by northeast- and northwest-striking lineaments	
References	Gröpper et al. (1991)	

10.2.6 Cripple Creek, Colorado, USA

Deposit	Cripple Creek
Mineralization	Epithermal gold
Production	About 590 tonnes gold
Location	Teller County, Colorado, USA
Host rocks	Syenite / Phonolite / Vogesite
Texture	Massive / Porphyritic / Porphyritic
Structure	Fine grained / Fine to medium grained / Medium grained
Mineralogy	Af-Pl-Qz±Bio / Phenocrysts: Foids-Ol-Cpx; Groundmass: Foids-Ol-Cpx / Phenocrysts: Hb; Groundmass: Af-Hb±Pl
Geochemistry	Alkaline, with high LILE and LREE / Alkaline, silica-undersaturated, with high LILE and LREE / Shoshonitic, with high LILE, moderate LREE, low HFSE
Ore minerals	Cal±Py±Cpy±Tet
Alteration	Potassic, propylitic, and sericitic
Age	Host rocks: 28-29 Ma; Mineralization: coeval with host rocks
Tectonic setting	Transitional between continental arc and within-plate setting
Structural features	Diatreme vent
References	Lindgren (1933), Walker and Walker (1956), Mutschler et al. (1985), Mutschler and Larson (1986), Kelly et al. (1998)

10.2.7   Dinkidi, Dipidio, Philippines

Deposit	Dinkidi
Mineralization	Porphyry copper-gold
Estimated reserves	120 million tonnes at 0.50 wt % Cu and 1.2 ppm Au
Location	Didipio Igneous Complex, North Luzon, Philippines
Source rocks	Monzodiorite	Monzonite
Texture	Massive	Porphyritic
Structure	Medium grained	Medium grained
Mineralogy	Pl-Qz-Cpx-Bio-Hb	Phenocrysts: Pl-Hb-Bio
		Groundmass: Pl-Af-Qz
Geochemistry		Shoshonitic, high LILE, moderate LREE, low HFSE
Ore minerals	Cpy±Bn
Alteration	Potassic, phyllic, and propylitic
Age	Host rocks: Early Miocene
	Mineralization: Early Miocene (23.2 ± 0.6 Ma), coeval with host rocks
Tectonic setting	Late oceanic arc
Structural features	Intersection of east-northeast-trending transfer structures with north-northwest-trending fractures parallel to the Philippine Fault
References	Wolfe et al. (1998), Wolfe (1999), R. Wolfe (written comm. 1999)

10.2.8   El Indio, Chile

Deposit	El Indio
Mineralization	Epithermal gold
Estimated reserves	About 140 tonnes gold, 770 tonnes silver and 306700 tonnes copper
Location	Chile, near border with Argentina; about 500 km north of Santiago and 180 km east of La Serena
Source rocks	Monzodiorite Granodiorite
Texture	Massive Massive
Structure	Medium grained Medium grained
Mineralogy	Qz-Pl-Af-Bio-Cpx Qz-Pl-Af-Bio-Hb-Cpx-Opx
Geochemistry	Potassic calc-alkaline, high LILE, moderate LREE, low HFSE
Ore minerals	En-Ten-Py± native gold
Alteration	Advanced argillic and pervasive silicification
Age	Host rocks: 16.7 Ma Mineralization: 16.0 Ma
Tectonic setting	Continental arc
Structural features	Northeast-striking faults dip 60-80° NW and define intensely veined area; probably related to caldera system
References	Walthier et al. (1985), Siddeley and Araneda (1986), Kay et al. (1987), Tschischow (1989), Jannas et al. (1990)

10.2.9   Emperor, Viti Levu, Fiji

Deposit	Emperor
Mineralization	Epithermal gold
Estimated reserves	About 150 tonnes gold
Location	Viti Levu, Fiji
Host rocks	Monzonite
Texture	Equigranular
Structure	Medium grained
Mineralogy	Pl-Af-Cpx-Phl
	Trachybasalt
	Porphyritic
	Medium grained
	Phenocrysts: Pl-Cpx-Ol±Phl±Hb
	Groundmass: Pl-Af-Cpx
Geochemistry	Shoshonitic, with high LILE, moderate LREE, low HFSE
Ore minerals	Py- gold-silver tellurides
Alteration	Propylitic and potassic
Age	Host rocks: Pliocene (4.3 Ma)
	Mineralization: Pliocene (3.71 ± 0.13 Ma)
Tectonic setting	Late oceanic arc
Structural features	Mineralization is located in caldera
References	Ahmad (1979), Anderson and Eaton (1990), Setterfield (1991), Setterfield et al. (1991, 1992), R. Jones (pers. comm., 1993)

10.2.10 Goonumbla, New South Wales, Australia

Deposit	Goonumbla		
Mineralization	Porphyry copper-gold, highest grades associated with potassic alteration		
Estimated reserves	About 166 million tonnes at 0.74 wt % Cu, 0.12 ppm Au, and 1.7 ppm Ag (about 1.2 million tonnes copper, 20 tonnes gold, and 280 tonnes silver)		
Location	New South Wales, Australia; 28 km northwest of Parkes		
Host rocks	Quartz monzonite	Latite	Trachyte
Texture	Equigranular	Porphyritic	Porphyritic
Structure	Coarse to medium grained	Coarse to medium grained	Medium grained
Mineralogy	Pl-Af-Qz-Cpx-Phl-Ap	Phenocrysts: Pl-Af-Cpx-Ap	Phenocrysts: Af-Pl-Cpx-Phl-Ap
		Groundmass: Qz-Af-Pl	Groundmass: Af-Pl-Qz
Geochemistry	Shoshonitic, with high LILE, moderate LREE, low HFSE		
Ore minerals	Cpy-Bn-Cc-Sl±Py		
Alteration	Propylitic and potassic		
Age	Host rocks: Ordovician (438 ± 3.5 Ma) Mineralization: Ordovician (439.2 ± 1.2 Ma), coeval with host rocks		
Tectonic setting	Late oceanic arc		
Structural features	Mineralization is located in caldera		
References	Heithersay et al. (1990), Perkins et al. (1990a, 1990b), Müller et al. (1994), Heithersay and Walshe (1995)		

10.2.11 Grasberg, Indonesia

Deposit	Grasberg
Mineralization	Porphyry copper-gold
Estimated reserves	About 362 million tonnes at 1.53 wt % Cu, 1.97 ppm Au, and 3.24 ppm Ag (about 5.5 million tonnes copper, 710 tonnes gold, and 1170 tonnes silver)
Location	Irian Jaya, Indonesia
Host rocks	Diorite
Texture	Porphyritic
Structure	Medium grained
Mineralogy	Phenocrysts: Pl-Cpx-Hb-Bio
	Groundmass: Af-Pl-Bio
Geochemistry	Highly potassic alkaline, with high LILE, low LREE, low HFSE, but relatively high Nb
Ore minerals	Cpy-Bn-Py±Cc±Dig±Cov
Alteration	Potassic, phyllic, and propylitic
Age	Host rocks: Pliocene (3.0 ± 0.5 Ma)
	Mineralization: within 1 Ma of igneous intrusion
Tectonic setting	Postcollisional arc (?)
Structural features	Mineralization controlled by intersections of reverse faults and sinistral strike-slip faults
References	Hickson (1991), Van Nort et al. (1991), MacDonald and Arnold (1994)

10.2.12   Kirkland Lake, Superior Province, Canada

Deposit	Kirkland Lake Mining District (example of Canadian lode-gold deposit)		
Mineralization	Mesothermal gold		
Estimated reserves	Not available (greater than 710 tonnes contained gold)		
Location	Superior Province, Ontario, Canada		
Spatially associated rocks	Syenite	Lamprophyre	
Texture	Equigranular	Porphyritic	
Structure	Medium grained	Medium grained	
Mineralogy	Af-Pl-Cpx-Bio-Mt±Ap	Phenocrysts: Hb±Phl	
		Groundmass: Pl-Af	
Geochemistry	Shoshonitic, with high LILE, moderate LREE, low HFSE		
Ore minerals	Cpy-Py		
Alteration	Secondary carbonate		
Age	Associated rocks: Archaean (2680 ± 1 Ma)		
	Mineralization: Archaean (2680 ± 1 Ma)		
Tectonic setting	Postcollisional arc		
Structural features	Mineralization and lamprophyres occur along major shear zones		
References	Wyman and Kerrich (1989a, 1989b), R. Kerrich (pers. comm., 1993), Rowins et al. (1993)		

10.2.13   Ladolam, Lihir Island, Papua New Guinea

Deposit	Ladolam	
Mineralization	Epithermal gold, highest grades associated with potassic alteration	
Estimated reserves	About 570 tonnes gold	
Location	Lihir Island, Papua New Guinea	
Host rocks	Monzonite	Latite	Trachybasalt
Texture	Equigranular	Porphyritic	Porphyritic
Structure	Fine to medium grained	Medium grained	Fine to medium grained
Mineralogy	Pl-Af-Cpx-Phl	Phenocrysts: Pl-Cpx±Phl±Hb	Phenocrysts: Pl-Cpx±Phl
		Groundmass: Af-Pl	Groundmass: Af-Pl
Geochemistry		Shoshonitic, with high LILE, moderate LREE, low HFSE	
Ore minerals	Py-Mar±Gn±Sl± gold-silver tellurides	
Alteration	Propylitic and potassic	
Age	Host rocks: Pleistocene (0.342–0.917 Ma)	
	Mineralization: Pleistocene (0.10–0.35 Ma)	
Tectonic setting	Late oceanic arc	
Structural features	Mineralization is located in caldera	
References	Wallace et al. (1983), Plimer et al. (1988), Moyle et al. (1990), Hoogvliet (1993), Spooner (1993)	

10.2.14 Maricunga Belt, Chile

Deposit	Maricunga Belt	
Mineralization	Porphyry copper-gold-silver, highest grades associated with potassic alteration	
Estimated reserves	About 420 tonnes gold and 14000 tonnes silver	
Location	Central Chile, between latitudes 26°00' and 28°00'	
Host rocks	Andesite	Diorite porphyry
Texture	Porphyritic	Porphyritic
Structure	Fine to medium grained	Fine to medium grained
Mineralogy	Phenocrysts: Pl-Hb-Bio±Af	Phenocrysts: Pl-Hb-Bio±Cpx
	Groundmass: Pl-Af-Qz	Groundmass: Qz-Pl-Af
Geochemistry	Potassic calc-alkaline, high LILE, moderate LREE, low HFSE	
Ore minerals	Py±Cpy±Bn±Mol±Sl±En±Ten	
Alteration	Potassic, propylitic, and argillic	
Age	Host rocks: Miocene (23–12.9 Ma)	
	Mineralization: Miocene (20–13 Ma)	
Tectonic setting	Continental arc	
Structural features	-	
References	Sillitoe (1991), Vila and Sillitoe (1991)	

10.2.15 Misima, Misima Island, Papua New Guinea

Deposit	Misima	
Mineralization	Epithermal gold	
Estimated reserves	About 56 million tonnes at 1.38 ppm Au and 21 ppm Au (about 75 tonnes gold and 1175 tonnes silver)	
Location	Island in the Louisiade Archipelago, Papua New Guinea; about 240 km east-southeast of the mainland	
Host rocks	Microgranodiorite	Spessartite
Texture	Porphyritic or phaneritic	Porphyritic
Structure	Coarse to medium grained	Fine to medium grained
Mineralogy	Phenocrysts: Pl-Qz-Bio-Hb Groundmass: Qz-Pl-Af	Phenocrysts: Hb Groundmass: Pl, Af
Geochemistry	Potassic calc-alkaline, with high LILE, moderate LREE, low HFSE	Shoshonitic, with high LILE, moderate LREE, low HFSE
Ore minerals	Py-Sl-Gn±Cpy± native gold ± native silver	
Alteration	Phyllic and propylitic	
Age	Host rocks: granodiorite Miocene (8.0 Ma); spessartite Pliocene (3.5 Ma) Mineralization: Pliocene (3.5 Ma)	
Tectonic setting	Postcollisional arc	
Structural features	Mineralization is controlled by major northwest-striking extensional fault	
References	Keyser (1961), Clarke (1988), Clarke et al. (1990), Lewis and Wilson (1990), Wilson and Barwick (1991), Appleby et al. (1995), A.-K. Appleby (written comm. 1996)	

10.2.16 Mount Kare, Papua New Guinea

Deposit	Mount Kare
Mineralization	Epithermal gold
Estimated reserves	Not determined
Location	Highlands of Papua New Guinea; about 18 km southwest of Porgera mine
Host rocks	Feldspar porphyry	Gabbro
Texture	Porphyritic	Equigranular
Structure	Coarse to medium grained	Medium grained
Mineralogy	Phenocrysts: Ol-Cpx-Hb-Phl-Ap	Ol-Cpx-Hb-Pl±Bio±An
	Groundmass: Af-Pl	
Geochemistry	Potassic alkaline, with high LILE, moderate LREE, very high Nb (up to 122 ppm)
Ore minerals	Py-Sl-Gn- gold-silver tellurides - native gold
Alteration	No potassic alteration
Age	Host rocks: Miocene (6.0 ± 0.1 Ma)
	Mineralization: Miocene (5.5 ± 0.1 Ma)
Tectonic setting	Postcollisional arc
Structural features	Mineralization occurs along major transform fault
References	Brunker and Caithness (1990), Richards and Ledlie (1993)

10.2.17   Mount Morgans, Eastern Goldfields, Western Australia

Deposit	Mount Morgans (example of Western Australia lode-gold deposit)	
Mineralization	Mesothermal gold	
Estimated reserves	About 12 tonnes gold	
Location	Eastern Goldfields Province, Yilgarn Block, Western Australia; between Leonora and Laverton	
Spatially associated rocks	Lamprophyre	Quartz-feldspar porphyry	Banded iron formation
Texture	Porphyritic	Porphyritic	Layered
Structure	Medium grained	Medium grained	Fine grained
Mineralogy	Phenocrysts: Phl-Cpx±Ap	Phenocrysts: Qz-Pl-Hb	Mt-Qz
	Groundmass: Pl-Af	Groundmass: Pl-Af	
Geochemistry	Potassic, with high LILE, moderate LREE, low HFSE		
Ore minerals	Py± native gold		
Alteration	Secondary carbonate		
Age	Associated rocks: Late Archaean (2684 ± 6 Ma)		
	Mineralization: Late Archaean (probably 2630 ± 10 Ma)		
Tectonic setting	Postcollisional arc		
Structural features	Mineralization, lamprophyres, and porphyries are controlled by fold and fault systems		
References	Perring (1988), Groves and Ho (1990), Vielreicher et al. (1994)		

10.2.18   Ok Tedi, Papua New Guinea

Deposit	Ok Tedi
Mineralization	Pophyry copper-gold
Estimated reserves	Oxide ore: 2.7 million tonnes at 2.08 ppm Au (about 5.5 tonnes gold); Sulphide ore: 355.3 million tonnes at 0.67 wt % Cu and 0.61 ppm Au (about 2.4 million tonnes copper and 215 tonnes gold)
Location	Northwestern part of Western Province, Papua New Guinea, near border with Indonesia
Host rocks	Quartz monzonite
Texture	Porphyritic
Structure	Fine to medium grained
Mineralogy	Phenocrysts: Pl-Af-Qz-Bio±Hb±Ap±Ti
	Groundmass: Pl-Af
Geochemistry	Potassic calc-alkaline, with high LILE, moderate LREE, low HFSE
Ore minerals	Py-Cpy±Mol±Bn± native gold
Alteration	Potassic, argillic, and propylitic
Age	Host rocks: Pliocene (2.6 Ma)
	Mineralization and alteration: Pleistocene (1.2-1.1 Ma)
Tectonic setting	Postcollisional arc
Structural features	Monzonite intrusion is controlled by a series of steeply dipping north-northwest- and north-northeast-striking faults
References	Bamford (1972), Arnold and Griffin (1978), Rush and Seegers (1990)

10.2.19 Porgera, Papua New Guinea

Deposit	Porgera
Mineralization	Epithermal gold
Estimated reserves	About 410 tonnes gold and 890 tonnes silver
Location	Highlands of Papua New Guinea
Host rocks	Trachybasalt	Feldspar porphyry
Texture	Porphyritic	Porphyritic
Structure	Medium grained	Coarse to medium grained
Mineralogy	Phenocrysts: Ol-Cpx-Pl-Phl-Ap	Phenocrysts: Ol-Cpx-Hb-Phl-Ap
	Groundmass: Af-Pl-Cpx	Groundmass: Af-Pl
Geochemistry	Highly potassic alkaline, with high LILE, moderate LREE, low HFSE, high Nb	
Ore minerals	Py- gold-silver tellurides - native gold
Alteration	No potassic alteration
Age	Host rocks: Miocene (6.0 ± 0.3 Ma)
	Mineralization: within 1 Ma of magmatism
Tectonic setting	Postcollisional arc
Structural features	Mineralization occurs along major transform fault
References	Handley and Henry (1990), Richards (1990a, 1990b, 1992), Richards et al. (1990, 1991), J. Standing (pers. comm., 1993)

10.2.20 Summitville, Colorado, USA

Deposit	Summitville
Mineralization	Epithermal gold
Production	About 20 tonnes gold
Location	San Juan volcanic field, Colorado, USA; about 10 km northwest of Platoro
Host rocks	Quartz latite Quartz monzonite
Texture	Porphyritic Porphyritic
Structure	Medium grained Medium grained
Mineralogy	Phenocrysts: Pl-Af-Qz-Bio-Hb±Ap Phenocrysts: Pl-Qz-Bio
	Groundmass: Af-Pl-Qz Groundmass: Af-Pl
Geochemistry	Potassic calc-alkaline, high LILE, moderate LREE, low HFSE
Ore minerals	En-Py±Cpy
Alteration	Advanced argillic
Age	Host rocks: Miocene (22.8 Ma)
	Mineralization: Miocene (22.4 Ma)
Tectonic setting	Continental arc
Structural features structure	Intrusions and mineralization are controlled by northwest-striking fault system and its intersection with a caldera ring
References	Mehnert et al. (1973), Stoffregen (1987), Gray and Coolbaugh (1994)

10.2.21   Tom's Gully, Northern Territory, Australia

Deposit	Tom's Gully
Mineralization	Mesothermal gold
Production	About 15 tonnes gold
Location	Mount Bundey area, Northern Territory, Australia
Spatially associated rocks	Syenite	Lamprophyre
Texture	Equigranular	Porphyritic
Structure	Coarse grained	Coarse to medium grained
Mineralogy	Af-Pl-Phl-Cpx-Ap	Phenocrysts: Ol-Phl-Hb
		Groundmass: Af-Hb-Phl±Ap±Ti
Geochemistry	Highly potassic, with high LILE, high LREE, high HFSE
Ore minerals	Py-Apy- native gold
Alteration	Secondary carbonate
Age	Associated rocks: Proterozoic (1831 ± 6 Ma)
	Mineralization: coeval with host rocks
Tectonic setting	Within plate
Structural features	Mineralization and lamprophyres are controlled by northeast-striking axial plane cleavage of metasedimentary rocks
References	Sheppard (1992, 1995), Sheppard and Taylor (1992), S. Sheppard and N.J. McNaughton (unpubl. ms, 1994)

10.2.22   Twin Buttes, Arizona, USA

Deposit	Twin Buttes
Mineralization	Porphyry copper, highest grades associated with potassic alteration
Production	About 1140 million tonnes at 0.7 wt % Cu (about 8 million tonnes copper)
Location	Pima County, Arizona, USA
Host rocks	Quartz-monzonite porphyry
Texture	Porphyritic
Structure	Coarse to medium grained
Mineralogy	Phenocrysts: Af-Bio±Ap
	Groundmass: Qz-Af-Pl
Geochemistry	Highly potassic
Ore minerals	Py-Cpy; as stockwork veining or disseminated
Alteration	Propylitic and potassic
Age	Host rocks: Paleocene (57.1 ± 2.1 Ma)
	Mineralization: coeval with host rocks
Tectonic setting	Continental arc
Structural features	Mineralization and intrusions are controlled by northeast-trending faults
References	Kelly (1977), Barter and Kelly (1982)

**10.2.23	Wiluna, Eastern Goldfields, Western Australia**

Deposit	Wiluna (example of Western Australia lode-gold deposits)
Mineralization	Mesothermal gold
Estimated reserves	About 105 tonnes gold
Location	Northern part of the Eastern Goldfields Province, Yilgarn Block, Western Australia
Spatially associated rocks	Lamprophyre Metabasalt
Texture	Porphyritic Porphyritic
Structure	Coarse grained Medium grained
Mineralogy	Phenocrysts: Hb-Af Phenocrysts: Hb-Pl-Cpx-Ol
	Groundmass: Qz-Pl-Af Groundmass: Pl-Cpx
Geochemistry	Lamprophyres are potassic, with high LILE, moderate LREE, low HFSE
Ore minerals	Py-Apy-Stib±Cpy±Sch± native gold
Alteration	Secondary carbonate
Age	Associated rocks: Late Archaean
	Mineralization: Late Archaean
Tectonic setting	Postcollisional arc
Structural features	Mineralization and lamprophyres are controlled by dextral strike-slip fault system
References	Hagemann et al. (1992), Hagemann (1993)

10.2.24   Woodlark Island, Papua New Guinea

Deposit	Woodlark Island	
Mineralization	Epithermal gold	
Estimated reserves	About 2.55 million tonnes at 3.7 ppm Au (about 9.5 tonnes gold)	
Location	In the Solomon Sea, Papua New Guinea; 600 km east of Prt Moresby and 160 km north of Misima Island	
Host rocks	Monzonite	Trachyandesite
Texture	Porphyritic	Porphyritic
Structure	Medium grained	Medium grained
Mineralogy	Phenocrysts: Pl-Bio	Phenocrysts: Pl-Hb
	Groundmass: Af-Pl	Groundmass: Af-Pl
Geochemistry	Shoshonitic, with high LILE, moderate LREE, low HFSE	Potassic calc-alkaline with high LILE, moderate LREE, low HFSE
Ore minerals	Py±Gn±Sl±Cpy±Tet	
Alteration	Phyllic and propylitic	
Age	Host rocks: Miocene (16.5-13 Ma) Mineralization: Miocene (12.3 Ma)	
Tectonic setting	Late oceanic arc	
Structural features	Mineralization and alteration are controlled by northwest-striking faults and a phreatic explosion breccia	
References	Russel and Finlayson (1987), Russell (1990), Corbett et al. (1994)	

References

Abdel-Rahman AM (1994) Nature of biotites from alkaline, calc-alkaline and peraluminous magmas. J Petrol 35: 525-541

Ahmad M (1979) Fluid inclusion and geochemical studies at the Emperor gold mine, Fiji. PhD Thesis, University of Tasmania, Hobart

Ahmad M, Walshe JL (1990) Wall-rock alteration at the Emperor gold-silver telluride deposit, Fiji. Austr J Earth Sci 37: 189-199

Alibert C, Michard A, Albarede F (1986) Isotope and trace element geochemistry of Colorado plateau volcanics. Geochim Cosmochim Acta 50: 2735-2750

Allan JF, Carmichael ISE (1984) Lamprophyric lavas in the Colima graben, SW Mexico. Contrib Mineral Petrol 88: 203-216

Allmendinger RW (1986) Tectonic development, southeastern border of the Puna Plateau, Northwestern Argentine Andes. Bull Am Geol Soc 97: 1070-1082

Allmendinger RW, Ramos VA, Jordan TE, Palma M, Isacks BL (1983) Paleogeography and Andean structural geometry, northwestern Argentina. Tectonics 2: 1-16

Anderson AT (1974) Chlorine, sulfur and water in magmas and oceans. Bull Am Geol Soc 85: 1485-1492

Anderson WB, Eaton PC (1990) Gold mineralization at the Emperor Mine, Vatukoula, Fiji. In: Hedenquist JW, White NC, Siddeley G (eds), Epithermal gold mineralization of the Circum-Pacific; geology, geochemistry, origin and exploration, II. J Geochem Expl 36: 267-296

Andrew RL (1995) Porphyry copper-gold deposits of the Southwest Pacific. Mining Engng 47 (1): 33-38

Aoki K, Ishiwaka K, Kanisawa S (1981) Fluorine geochemistry of basaltic rocks from continental and oceanic regions and petrogenetic application. Contrib Mineral Petrol 76: 53-59

Appleby AK, Circosta G, Fanning M, Logan K (1995) A new model for controls on Au-Ag mineralization on Misima Island, Papua New Guinea. In: Proceedings of 101st Annual Northwest Mining Association Convention. Northwest Mining Association, Spokane, pp 122-127

Appleton JD (1972) Petrogenesis of potassium-rich lavas from the Roccamonfina volcano, Roman Region, Italy. J Petrol 13: 425-456

Arculus RJ (1985) Oxidation status of the mantle: past and present. Ann Rev Earth Planet Sci 13: 75-95

Arculus RJ, Johnson RW (1978) Criticism of generalised models for the magmatic evolution of arc-trench systems. Earth Planet Sci Lett 39: 118-126

Arnold GO, Griffin TJ (1978) Intrusions and porphyry copper prospects of the Star Mountains, Papua New Guinea. Econ Geol 73: 785-795

Ashley PM, Cook NDJ, Hill RL, Kent AJR (1994) Shoshonitic lamprophyre dykes and their relation to mesothermal Au-Sb veins at Hillgrove, New South Wales, Australia. Lithos 32: 249-272

Bacon CA (1990) Calc-alkaline, shoshonitic and primitive tholeiitic lavas from monogenetic volcanoes near Crater Lake, Oregon. J Petrol 31: 135-166

Bailey DK (1982) Mantle metasomatism — continuing chemical change within the Earth. Nature 296: 525-530

Bailey DK, Hampton CM (1990) Volatiles in alkaline magmatism. Lithos 26: 157-165

Bailey JC, Frolova TI, Burikova IA (1989) Mineralogy, geochemistry and petrogenesis of Kurile island-arc basalts. Contrib Mineral Petrol 102: 265-280

Baker N, Fryer P, Martinez F, Yamazaki T (1996) Rifting history of the northern Mariana Trough: SeaMARC II and seismic reflection surveys. J Geophys Res 101: 11427-11455

Ballhaus C (1993) Redox states of lithospheric and asthenospheric upper mantle. Contrib Mineral Petrol 114: 331-348

Ballhaus C (1995) Is the upper mantle metal-saturated? Earth Planet Sci Lett 132: 75-86

Ballhaus C, Berry RF, Green DH (1990) Oxygen fugacity controls in the Earth's upper mantle. Nature 348: 437-440

Bamford RW (1972) The Mount Fubilan (Ok Tedi) porphyry copper deposit, Territory of Papua and New Guinea. Econ Geol 67: 1019-1033

Barker DS (1983) Igneous rocks. Prentice Hall, Englewood Cliffs, NJ: 417 pp

Barley ME, Groves DI (1990) Deciphering the tectonic evolution of Archean greenstone belts: the importance of contrasting histories to the distribution of mineralization in the Yilgarn Craton, Western Australia. Precambr Res 46: 3-20

Barley ME, Groves DI (1992) Supercontinent cycles and the distribution of metal deposits through time. Geology 20: 291-294

Barley ME, Eisenlohr BN, Groves DI, Perring CS, Vearncombe JR (1989) Late Archean convergent margin tectonics and gold mineralization: a new look at the Norseman-Wiluna Belt, Western Australia. Geology 17: 826-829

Barrie CT (1993) Petrochemistry of shoshonitic rocks associated with porphyry copper-gold deposits of central Quesnellia, British Columbia, Canada. J Geochem Expl 48: 225-258

Barter CF, Kelly JL (1982) Geology of the Twin Buttes mineral deposit, Pima Mining district, Pima County, Arizona. In: Titley SR (ed) Advances in geology of the porphyry copper deposits, Southwestern North America. University of Arizona Press, Tucson, Arizona, pp 407-432

Barton M (1979) A comparative study of some minerals occurring in the potassium-rich alkaline rocks of the Leucite Hills, Wyoming, the Vico volcano, western Italy, and the Toro Ankole region, Uganda. Neues Jb Mineral Abh 137: 113-134

Barton M, van Bergen MJ (1981) Green clinopyroxenes and associated phases in a potassium-rich lava from the Leucite Hills, Wyoming. Contrib Mineral Petrol 77: 101-114

Beccaluva L, Bigioggero LB, Chiesa S, Colombo A, Fanti G, Gatto GO, Gregnanin A, Montrasio A, Piccirillo EM, Tunesi A (1983) Postcollisional orogenic dyke magmatism in the Alps. Mem Soc Geol Ital 26: 341-359

Bergman SC (1987) Lamproites and other potassium-rich igneous rocks: a review of their occurrence, mineralogy and geochemistry. In: Fitton JG, Upton BGJ (eds) Alkaline igneous rocks. Geological Society, London, pp 103-190 (Geol Soc Spec Publ 30)

Bergman SC, Dunn DP, Krol LG (1988) Rock and mineral chemistry of the Linhaisai minette and the origin of Borneo diamonds, Central Kalimantan, Indonesia. Can Mineral 26: 23-44

Blevin PL, Chappell BW (1996) Controls on the distribution and character of the intrusive-metallogenic provinces of Eastern Australia. Geol Soc Austr Abst 41: 42

Bloomer SH, Stern RJ, Fisk E, Geschwind CH (1989) Shoshonitic volcanism in the Northern Marianas arc. 1. Mineralogic and major and trace element characteristics. J Geophys Res 94: 4469-4496

Bonham HF, Giles DL (1983) Epithermal gold/silver deposits: the geothermal connection. Geotherm Resources Council Spec Rep 13: 257-262

Boudreau AE, Mathez EA, McCallum IS (1986) Halogen geochemistry of the Stillwater and Bushveld complexes: evidence for transport of the platinum-group elements by Cl-rich fluids. J Petrol 27: 967-986

Bowman JR, Parry WT, Kropp WP, Kruer SA (1987) Chemical and isotopic evolution of hydrothermal solutions at Bingham, Utah. Econ Geol 82: 395-428

Briqueu L, Bougault H, Joron JL (1984) Quantification of Nb, Ta, Ti and V anomalies in magmas associated with subduction zones: petrogenetic implications. Earth Planet Sci Lett 68: 297-308

Brooks C, Ludden J, Pigeon Y, Hubregtse JJMW (1982) Volcanism of shoshonite to high-K andesite affinity in an Archean arc environment, Oxford Lake, Manitoba. Can J Earth Sci 19: 55-67

Brown GC, Thorpe RS, Webb PC (1984) The geochemical characteristics of granitoids in contrasting arcs and comments on magma sources. J Geol Soc 141: 413-426

Brügmann GE, Arndt NT, Hofmann AW, Tobschall HJ (1987) Noble metal abundances in komatiite suites from Alexo, Ontario, and Gorgona Island, Colombia. Geochim Cosmochim Acta 51: 2159-2169

Brunker RL, Caithness SJ (1990) Mount Kare gold deposit. In: Hughes FE (ed) Geology of the mineral resources of Australia and Papua New Guinea. The Australasian Institute of Mining and Metallurgy, Parkville, pp 1755-1758 (Australas Inst Min Metall Monogr 14)

Burrows DR, Spooner ETC (1989) Relationships between Archean gold quartz vein-shear zone mineralization and igneous intrusions in the Val d'Or and Timmins areas, Abitibi Subprovince, Canada. In: Keays RR, Ramsay WRH, Groves DI (eds), The geology of gold deposits: the perspective in 1988. The Economic Geology Publishing Company, El Paso, pp 424-444 (Econ Geol Monogr 6)

Burrows DR, Spooner ETC (1991) The Lamaque Archean stockwork Au system, Val d'Or, Quebec: relationship to a ≈20 Ma, calc-alkaline, intrusive sequence. GAC-MAC Prog with Abst 16: 17

Butler BS (1920) Oquirrh range. In: The ore deposits of Utah. US Geological Survey, pp 335-362 (US Geol Surv Prof Paper 111)

Cabri LJ (1981) Platinum-group elements: mineralogy, geology, recovery. Can Inst Min Metall Spec Vol 23: 267 pp

Caelles JC (1979) The geological evolution of the Sierras Pampeanas Massif, La Rioja and Catamarea Provinces, Argentina. PhD Thesis, Queen's University, Kingston

Campbell IH, Naldrett AJ, Barnes SJ (1983) A model for the origin of the platinum-rich sulphide horizons in the Bushveld and Stillwater complexes. J Petrol 24: 133-165

Campbell IH, Compston DM, Richards JP, Johnson JP, Kent AJR (1998) Review of the application of isotopic studies to the genesis of Cu-Au mineralisation at Olympic Dam and Au mineralisation at Porgera, the Tennant Creek district and Yilgarn Craton. Austr J Earth Sci 45: 201-218

Candela PA (1989) Felsic magmas, volatiles, and metallogenesis. In: Whitney JA, Naldrett AJ (eds) Ore deposition associated with magmas. The Economic Geology Publishing Company, El Paso, pp 223-233 (Rev Econ Geol 4)

Candela PA (1997) A review of shallow, ore-related granites: textures, volatiles, and ore metals. J Petrol 38: 1619-1633

Carmichael ISE, Ghiorso MS (1986) Oxidation-reduction relations in basic magma: a case for homogeneous equilibria. Earth Planet Sci Lett 78: 200-210

Carmichael ISE, Lange RA, Luhr JF (1996) Quaternary minettes and associated volcanic rocks of Mascota, western Mexico: a consequence of plate extension above a subduction modified mantle wedge. Contrib Mineral Petrol 124: 302 333

Carr PF (1998) Subduction-related Late Permian shoshonites of the Sydney Basin, Australia. Mineral Petrol 63: 49-71

Carten RB (1987) Evolution of immiscible Cl- and F-rich liquids from ore magmas, Henderson porphyry molybdenum deposit, Colorado. Geol Soc Am Abst 19: 613

Civetta L, Innocenti F, Manetti P, Peccerillo A, Poli G (1981) Geochemical characteristics of potassic volcanics from Mts. Ernici, southern Latium, Italy. Contrib Mineral Petrol 78: 37-47

Clark AH (1993) Are outsize porphyry copper deposits either anatomically or environmentally distinctive? In: Whiting BH, Hodgson CJ, Mason R (eds) Giant ore deposits. The Economic Geology Publishing Company, El Paso, pp 213-283 (Econ Geol Spec Publ 2)

Clark DS (1988) Lateral zonation within epithermal gold mineralization, Misima Island, Papua New Guinea. Australas Inst Min Metall Bull Proc 293: 63-66

Clark DS, Lewis RW, Waldron HM (1990) Geology and trace-element geochemistry of the Umuna gold-silver deposit, Misima Island, Papua New Guinea. In: Hedenquist JW, White NC, Siddeley G (eds), Epithermal gold mineralization of the Circum-Pacific; geology, geochemistry, origin and exploration, II. J Geochem Expl 36: 201-223

Cline JS, Bodnar RJ (1991) Can economic porphyry copper mineralization be generated by a typical calc-alkaline melt? J Geophys Res 96: 8113-8126

Cocco G, Fanfani L, Zanazzi PF (1972) Potassium. In: Wedepohl KH (ed) Handbook of geochemistry, vol II-2. Springer-Verlag, Berlin.

Coleman PJ (1970) Geology of the Solomon and New Hebrides islands as part of the Melanesian re-entrant, southwest Pacific. Pac Sci 24: 289-314

Colley H, Greenbaum D (1980) The mineral deposits and metallogenesis of the Fiji platform. Econ Geol 75: 807-829

Colvine AC (1989) An empirical model for the formation of Archaean gold deposits: products of final cratonization of the Superior Province, Canada. In: Keays RR, Ramsay WRH, Groves DI (eds), The geology of gold deposits: the perspective in 1988. The Economic Geology Publishing Company, El Paso, pp 37-53 (Econ Geol Monogr 6)

Conticelli S, Peccerillo A (1992) Petrology and geochemistry of potassic and ultrapotassic volcanism in central Italy: petrogenesis and inferences on the evolution of the mantle sources. Lithos 28: 221-240

Contini S, Venturelli G, Toscani L, Capedri S, Barbieri M (1993) Cr-Zr-armalcolite-bearing lamproites of Cancarix, SE Spain. Mineral Mag 57: 203-216

Cooke DR, McPhail DC, Bloom MS (1996) Epithermal gold mineralization, Acupan, Baguio district, Philippines: geology, mineralization, alteration, and the thermochemical environment of ore deposition. Econ Geol 91, 243-272

Cooper JR (1973) Geologic map of the Twin Buttes quandrangle southwest of Tucson, Pima County, Arizona. US Geol Surv Misc Geol Inv Map I-745

Cooper P, Taylor B (1987) Seismotectonics of New Guinea: a model for arc reversal following arc-continent collision. Tectonics 6: 53-67

Corbett GJ, Leach TM (1998) Southwest Pacific Rim and gold-copper systems: structure, alteration, and mineralization. Econ Geol Spec Publ 6: 236 pp

Corbett GJ, Leach TM, Shatwell DO, Hayward SB (1994) Gold mineralization on Woodlark Island, Papua New Guinea. In: Rogerson R (ed) Geology, exploration and mining conference, June 1994, Lae, Papua New Guinea, proceedings. Australasian Institute of Mining and Metallurgy, Parkville, pp 92-100

Cox KG, Hawkesworth CJ, O'Nions RK (1976) Isotopic evidence for the derivation of some Roman Region volcanics from anomalously enriched mantle. Contrib Mineral Petrol 56: 173-180

Crawford AJ, Corbett KD, Everard JL (1992) Geochemistry of the Cambrian volcanic-hosted massive sulfide-rich Mount Read Volcanics, Tasmania, and some tectonic implications. Econ Geol 87: 597-619

Creaser RA (1996) Petrogenesis of a Mesoproterozoic quartz latite-granitoid suite from the Roxby Downs area, South Australia. Precambr Res 79: 371-394

Crocket JH (1979) Platinum-group elements in mafic and ultramafic rocks: a survey. Can Mineral 17: 391-402

Cundari A (1973) Petrology of the leucite-bearing lavas in New South Wales. J Geol Soc Austr 20: 465-492

Cundari A (1979) Petrogenesis of leucite-bearing lavas in the Roman volcanic region, Italy. The Sabatini lavas. Contrib Mineral Petrol 70: 9-21

Cundari A, Le Maitre RW (1970) On the petrogenesis of the leucite-bearing rocks of the Roman and Birunga volcanic regions. J Petrol 11: 33-47

Cundari A, Matthias PP (1974) Evolution of the Vico lavas, Roman volcanic region, Italy. Bull Volcanol 38: 98-114

Curray JR, Shor GG, Raitt RW, Henry M (1977) Seismic refraction and reflection studies of crustal structure of the eastern Sunda and western Banda arcs. J Geophys Res 82: 2479-2489

Dal Piaz GV, Venturelli G, Scolari A (1979) Calc-alkaline to ultrapotassic postcollisional volcanic activity in the internal northwestern Alps. Mem Inst Geol Mineral, Univ Padova 32: 4-16

Daly RA (1910) Origin of the alkaline rocks. Bull Geol Soc Am 21: 87-115

Dalziel IWD (1995) Earth before Pangea. Scient Amer 272: 58-63

D'Antonio M, Di Girolamo P (1994) Petrological and geochemical study of mafic shoshonitic volcanics from Procida-Vivara and Ventotene Islands (Campanian Region, South Italy). Acta Vulcanol 5: 69-80

Davidson JP, Mpodozis C (1991) Regional geologic setting of epithermal gold deposits, Chile. Econ Geol 86: 1174-1186

Davidson JP, McMillan NJ, Moorbath S, Wörner G, Harmon RS, Lopez-Escobar L (1990) The Nevados de Payachata volcanic region (18° S/69° W, N. Chile) II. Evidence for widespread crustal involvement in Andean magmatism. Contrib Mineral Petrol 105: 412-432

Dawson JB (1987) The kimberlite clan: relationship with olivine and leucite lamproites, and inferences for upper-mantle metasomatism. In: Fitton JG, Upton BGJ (eds) Alkaline igneous rocks. Geological Society, London, pp 95-101 (Geol Soc Spec Publ 30)

De Long SE, Hodges FN, Arculus RJ (1975) Ultramafic and mafic inclusions, Kanaga Island, Alaska, and the occurrence of alkaline rocks in island arcs. J Geol 83: 721-736

Demant A, Lestrade P, Lubala RT, Kampunzu AB, Durieux J (1994) Volcanological and petrological evolution of Nyiragongo volcano, Virunga volcanic field, Zaire. Bull Volcanol 56: 47-61

De Mulder M, Hertogen J, Deutsch S, Andre L (1986) The role of crustal contamination in the potassic suite of the Karisimbi Volcano (Virunga, African Rift Valley). Chem Geol 57: 117-136

De Paolo DJ, Johnson RW (1979) Magma genesis in the New Britain island arc: constraints from Nd and Sr isotopes and trace element patterns. Contrib Mineral Petrol 70: 367-379

Deutsch A (1980) Alkalibasaltische Ganggesteine aus der westlichen Goldeckgruppe (Kärnten/ Österreich). Tschermaks Min Petr Mitt 27: 17-34

Deutsch A (1984) Young Alpine dykes south of the Tauern Window (Austria): a K-Ar and Sr isotope study. Contrib Mineral Petrol 85: 45-57

Deutsch A (1986) Geochemie oligozäner shoshonitischer Ganggesteine aus der Kreuzeckgruppe (Kärnten/Osttirol). Mitt Ges Geol Bergbaustud Österr 32: 105-124

De Wit MJ (1989) Book review: alkaline igneous rocks. Lithos 24: 81-82

Di Girolamo P (1984) Magmatic character and geotectonic setting of some Tertiary-Quaternary Italian volcanic rocks: orogenic, anorogenic and transitional association — a review. Bull Volcanol 47: 421-432

Dimock RR, Carter CJ, Forth JR, Hoogvliet H, Jacobsen WL, Ketcham VJ, Probert TI, Thiel KF, Watson DL, Zavodni ZM (1993) Proposed gold ore mining and treatment operations at the Lihir Island Project, New Ireland Province, Papua New Guinea. Australas Inst Min Metall Monogr 19: 913-922

Dixon TH, Batiza R (1979) Petrology and chemistry of Recent lavas in the N-Marianas: implications for the origin of island-arc basalts. Contrib Mineral Petrol 70: 167-181

Dixon JE, Clague DA, Wallace P, Poreda R (1997) Volatiles in alkalic basalts from the North Arch Volcanic Field, Hawaii: extensive degassing of deep submarine-erupted alkalic series lavas. J Petrol 38: 911-939

Dodge FCW, Moore JG (1981) Late Cenozoic volcanic rocks of the southern Sierra Nevada, California: II. Geochemistry. Bull Geol Soc Am 92: 1670-1761

Dostal J, Dupuy C, Lefevre C (1977a) Rare earth element distribution in Plio-Quaternary volcanic rocks from southern Peru. Lithos 10: 173-183

Dostal J, Zentilli M, Caelles JC, Clark AH (1977b) Geochemistry and origin of volcanic rocks of the Andes (26-28° S). Contrib Mineral Petrol 63: 113-128

Duggan MB, Jaques AL (1996) Mineralogy and geochemistry of Proterozoic shoshonitic lamprophyres from the Tennant Creek Inlier, Northern Territory. Austr J Earth Sci 43: 269-278

Duncker KE, Wolff JA, Harmon RS, Leat PT, Dickin AP, Thompson RN (1991) Diverse mantle and crustal components in lavas of the NW Cerros del Rio volcanic field, Rio Grande Rift, New Mexico. Contrib Mineral Petrol 108: 331-345

Dunham KC (1974) Geological setting of the useful minerals in Britain. Proceed R Soc A339: 273-288

Edgar AD (1980) Role of subduction in the genesis of leucite-bearing rocks: discussion. Contrib Mineral Petrol 73: 429-431

Edgar AD (1987) The genesis of alkaline magmas with emphasis on their source regions: inferences from experimental studies. In: Fitton JG, Upton BGJ (eds) Alkaline igneous rocks. Geological Society, London, pp 29-52 (Geol Soc Spec Publ 30)

Edgar AD, Arima M (1985) Fluorine and chlorine contents of phlogopites crystallized from ultrapotassic rock compositions in high pressure experiments: implications for halogen reservoirs in source regions. Am Mineral 70: 529-536

Edgar AD, Charbonneau HE (1991) Fluorine-bearing phases in lamproites. Mineral Petrol 44: 125-149

Edgar AD, Mitchell RH (1997) Ultra high pressure-temperature melting experiments on an SiO_2-rich lamproite from Smoky Butte, Montana: derivation of siliceous lamproite magmas from enriched sources deep in the continental mantle. J Petrol 38: 457-477

Edgar AD, Lloyd FE, Vukadinovic D (1994) The role of fluorine in the evolution of ultrapotassic magmas. Mineral Petrol 51: 173-193

Edwards CMH, Menzies MA, Thirlwall MF, Morrid JD, Leeman WP, Harmon RS (1994) The transition to potassic alkaline volcanism in island arcs: the Ringgit-Beser Complex, East Java, Indonesia. J Petrol 35: 1557-1595

Einaudi MT (1982) Description of skarns associated with porphyry copper plutons. In: Titley SR (Editor) Advances in geology of the porphyry copper deposits, Southwestern North America. University of Arizona Press, Tucson, pp 139-183

Ellam RM, Hawkesworth CJ (1988) Elemental and isotopic variations in subduction related basalts: evidence for a three component model. Contrib Mineral Petrol 98: 72-80

Ellam RM, Hawkesworth CJ, Menzies MA, Rogers NW (1989) The volcanism of Southern Italy: role of subduction and the relationship between potassic and sodic alkaline magmatism. J Geophys Res 94: 4589-4601

Exner C (1976) Die geologische Position der Magmatite des Periadriatischen Lineamentes. Verh Geol Bundesanstalat Wien Austria 1976: 3-64

Fears DW, Mutschler FE, Larson EE (1986) Cripple Creek, Colorado — a petrogenetic model. In: Abstracts with Program, Geological Society of America 99th Annual Meeting, San Antonio, p 599 (Geol Soc Am, Abst with Prog 18)

Feldstein SN, Lange RA, Vennemann T, O'Neil JR (1996) Ferric-ferrous ratios, H_2O contents and D/H ratios of phlogopite and biotite from lavas of different tectonic regimes. Contrib Mineral Petrol 126: 51-66

Ferguson AK, Cundari A (1975) Petrological aspects and evolution of the leucite-bearing lavas from Bufumbira, SW-Uganda. Contrib Mineral Petrol 50: 25-46

Finlayson EJ, Rock NMS, Golding SD (1988) Deformation and regional carbonate metasomatism of turbidite-hosted Cretaceous alkaline lamprophyres, northwestern Papua New Guinea. Chem Geol 69: 215-233

Fleming AW, Handley GA, Williams KL, Hills AL, Corbett GJ (1986) The Porgera gold deposit, Papua New Guinea. Econ Geol 81: 660-680

Floyd PA, Winchester JA (1975) Magma type and tectonic setting discrimination using immobile elements. Earth Planet Sci Lett 27: 211-218

Foden JD (1979) The petrology of some young volcanic rocks from Lombok and Sumbawa, Lesser Sunda islands. PhD Thesis, University of Tasmania, Hobart

Foden JD, Varne R (1980) The petrology and tectonic setting of Quaternary-Recent volcanic centres of Lombok and Sumbawa, Sunda arc. Chem Geol 30: 201-226

Foley SF (1984) Liquid immiscibility and melt segregation in alkaline lamprophyres from Labrador. Lithos 17: 127-137

Foley SF (1992) Petrological characterization of the source components of potassic magmas: geochemical and experimental constraints. Lithos 28: 187-204

Foley SF (1994) Geochemische und experimentelle Untersuchungen zur Genese der kalireichen Magmatite. Neues Jb Mineral Abh 167: 1-55

Foley SF, Peccerillo A (1992) Potassic and ultrapotassic magmas and their origin. Lithos 28: 181-185

Foley SF, Wheller GE (1990) Parallels in the origin of the geochemical signatures of island arc volcanics and continental potassic igneous rocks: the role of residual titanates. Chem Geol 85: 1-18

Foley SF, Venturelli G, Green DH, Toscani L (1987) The ultrapotassic rocks: characteristics, classification, and constraints for petrogenetic models. Earth Sci Rev 24: 81-134

Fornaseri M, Scherillo A, Ventriglia U (1963) La regione vulcanica dei Colli Albani; vulcano Laziale. Consiglio Naz Ricerche, Centro Miner e Petrog, Rome, 561 pp

Friedrich OM (1963) Die Lagerstätten der Kreuzeck-Gruppe. Monographien Kärntner Lag 3. Teil Archiv für Lagerstättenforschung in den Ostalpen, Band 1, Mining University, Leoben

Fuge R, Andrews MJ, Johnson CC (1986) Chlorine and iodine, potential pathfinder elements in exploration geochemistry. Appl Geochem 1: 111-116

Gallo F, Giammetti F, Venturelli G, Vernia L (1984) The kamafugitic rocks of San Venanzo and Cuppaello, Central Italy. Neues Jb Mineral Mh 5: 198-210

Gammons CH, Williams-Jones AE (1997) Chemical mobility of gold in the porphyry-epithermal environment. Econ Geol 92: 45-59

Garcia MO, Liu NWK, Muenow DW (1979) Volatiles in submarine volcanic rocks from the Mariana island arc and trough. Geochim Cosmochim Acta 43: 305-312

Garrett S (1996) The geology and mineralization of the Dinkidi porphyry-related Au-Cu deposit. In: Porphyry-related copper and gold deposits of the Asia Pacific region. Australian Mineral Foundation, Cairns, pp 6.0-6.15

Gee RD, Baxter JL, Wilde SA, Williams IR (1981) Crustal development in the Archaean Yilgarn Block, Western Australia. In: Glover JE, Groves DI (eds) Archaean geology: second international symposium, Perth, 1980. Geological Society of Australia, Sydney, pp 43-56 (Geol Soc Austr Spec Publ 7)

Gest DE, McBirney AR (1979) Genetic relationships of shoshonitic and absarokitic magmas, Absaroka Mountains, Wyoming. J Volcanol Geotherm Res 6: 85-104

Ghiara MR, Lirer L (1976) Mineralogy and geochemistry of the "low potassium" series of the Roccamonfina volcanic suite (Campania, S-Italy). Bull Volcanol 40: 39-56

Gibson SA, Thompson RN, Leat PT, Morrison MA, Hendry GL, Dickin AP, Mitchell JG (1993) Ultrapotassic magmas along the flanks of the Oligo-Miocene Rio Grande Rift, U.S.A.: monitors of the zone of lithospheric mantle extension and thinning beneath a continental rift. J Petrol 34: 187-228

Gill JB (1970) Geochemistry of Viti Levu, Fiji, and its evolution as an island arc. Contrib Mineral Petrol 27: 179-203

Gill JB, Whelan P (1989) Early rifting of an oceanic island arc (Fiji) produced shoshonitic to tholeiitic basalts. J Geophys Res 94: 4561-4578

Glasmacher U, Günther F (1991) Gold-bearing sulfide veins in shoshonites formed by high-T, high-Cl alkaline fluids, Prospector Mtn, Yukon Territory. GAC/MAC Prog with Abst 16: A46

Gonzalez OE (1975) Geologia y alteracion en el cobre porfidico "Bajo La Alumbrera", Republica Argentina. II Congreso Ibero-Americano de Geologia Economica 2: 247-270

Gorton MP (1977) The geochemistry and origin of Quaternary volcanism in the New Hebrides. Geochim Cosmochim Acta 41: 1257-1270

Götze HJ, Lahmeyer B, Schmidt S, Strunk S (1987) Gravity field and megafault-system of the Central Andes (20°-26° S). Terra Cognita 7: 57

Gray JE, Coolbaugh MF (1994) Geology and geochemistry of Summitville, Colorado: an epithermal acid sulfate deposit in a volcanic dome. Econ Geol 89: 1906-1923

Greenough JD, Hayatsu A, Papezik VS (1988) Mineralogy, petrology and geochemistry of the alkaline Malpeque sill, Prince Edward Island. Can Mineral 26: 97-108

Gribble RF, Stern RJ, Newman S, Bloomer SH, O'Hearn T (1998) Chemical and isotopic composition of lavas from the northern Mariana Trough: implications for magmagenesis in back-arc basins. J Petrol 39: 125-154

Gröpper H, Calvo M, Crespo H, Bisso CR, Cuadra WA, Dunkerley PM, Aguirre E (1991) The epithermal gold-silver deposit of Choquelimpie, Northern Chile. Econ Geol 86: 1206-1221

Groves DI (1982) The Archean and earliest Proterozoic evolution and metallogeny of Australia. Rev Brasil Geocienc 12: 135-148

Groves DI (1993) The crustal continuum model for late-Archaean lode-gold deposits of the Yilgarn Block, Western Australia. Mineralium Deposita 28: 366-374

Groves DI, Ho SE (1990) A short review of gold in the Yilgarn Block. In: Hughes FE (ed) Geology of the mineral resources of Australia and Papua New Guinea. The Australasian Institute of Mining and Metallurgy, Parkville, pp 539-553 (Australas Inst Min Metall Monogr 14)

Groves DI, Barley ME, Shepherd JM (1994) Geology and mineralisation of Western Australia. In: Dentith MC, Frankcombe KF, Ho SE, Shepherd JM, Groves DI, Trench A (eds) Geophysical signatures of Western Australian mineral deposits. Geology Key Centre & UWA Extension, The University of Western Australia, and Australian Society of Exploration Geophysicists, Perth, pp 3-28 (Geol Geophys Dept & UWA Extension, Univ West Austr, Publ 26)

Gueddari K, Piboule M, Amosse J (1996) Differentiation of platinum-group elements (PGE) and of gold during partial melting of peridotites in the lherzolitic massifs of the Betico-Rifean Range (Ronda and Beni Bousera). Chem Geol 134: 181-197

Guilbert JM (1995) Geology, alteration, mineralization, and genesis of the Bajo de la Alumbrera porphyry copper-gold deposit, Catamerca Province, Argentina. In: Pierce FW, Bolm JG (eds) Porphyry copper deposits of the America Cordillera. Arizona Geol Soc Digest 20: 646-656

Gunow AJ, Ludington S, Munoz JL (1980) Fluorine in micas from the Henderson molybdenite deposit, Colorado. Econ Geol 75: 1127-1136

Gustafson LB, Hunt JP (1975) The porphyry copper deposit at El Salvador, Chile. Econ Geol 70: 857-912

Hagemann SG (1993) The Wiluna lode-gold deposits, Western Australia: a case study of a high crustal level Archean lode-gold system. PhD Thesis, The University of Western Australia, Perth

Hagemann SG, Groves DI, Ridley JR, Vearncombe JR (1992) The Archean lode gold deposits at Wiluna, Western Australia: high-level brittle-style mineralization in a strike-slip regime. Econ Geol 87: 1022-1053

Haggerty SE (1990) Redox state of the continental lithosphere. In: Menzies MA (ed) Continental mantle. Clarendon Press, Oxford, pp 87-109

Haggman JA (1997a) Philippines exploration: discovery of the Dinkidi gold-copper porphyry. Mining Engng 49 (3): 34-39

Haggman JA (1997b) The discovery, exploration and future development of the Didipio Project. In: World Gold '97 Conference. The Australasian Institute of Mining and Metallurgy, Publ Ser 2/97: 183-188

Hallberg JA (1985) Geology and mineral deposits of the Leonora-Laverton area, northeastern Yilgarn Block, Western Australia. Hesperian Press, Perth: 89 pp

Halliday AN, Davidson JP, Holden P, DeWolf C, Lee DC, Fitton JG (1990) Trace element fractionation in plumes and the origin of HIMU mantle beneath the Cameroon Line. Nature 347: 523-528

Handley GA, Henry DD (1990) Porgera gold deposit In: Hughes FE (ed) Geology of the mineral deposits of Australia and Papua New Guinea. The Australasian Institute of Mining and Metallurgy, Parkville, pp 1717-1724 (Australas Inst Min Metall Monogr 14)

Harris PG (1957) Zone refining and the origin of potassic basalts. Geochim Cosmochim Acta 12: 195-208

Harte B, Hawkesworth CJ (1989) Mantle domains and mantle xenoliths. In: Ross J, Jaques AL, Ferguson J, Green DH, O'Reilly SY, Danchin RV, Janse AJA (eds) Kimberlites and related rocks. Geological Society of Australia, Sydney, pp 649-686 (Geol Soc Austr Spec Publ 14)

Hatherton T, Dickinson WR (1969) The relationship between andesite volcanism and seismicity in Indonesia, the Lesser Antilles and other island arcs. J Geophys Res 74: 5301-5310

Hayashi KI, Ohmoto H (1991) Solubility of gold in NaCl- and H_2S-bearing aqueous solutions at 250-350°C. Geochim Cosmochim Acta 55: 2111-2126

Hayba DO, Bethke PM, Heald P, Foley NK (1985) Geologic, mineralogic and geochemical characteristics of volcanic-hosted precious-metal deposits. In: Berger BR, Bethke PM (eds) Geology and geochemistry of epithermal systems. The Economic Geology Publishing Company, El Paso, pp 129-167 (Rev Econ Geol 2)

Heald P, Foley NK, Hayba DO (1987) Comparative anatomy of volcanic-hosted epithermal deposits: acid-sulfate and adularia-sericite types. Econ Geol 82: 1-24

Heithersay PS (1986) Endeavour 26 North copper-gold deposit, Goonumbla, N.S.W. — paragenesis and alteration zonation. In: Berkman DA (ed) 13th CMMI congress, volume 2, geology and exploration. 13th Congress of the Council of Mining and Metallurgical Institutions, and The Australasian Institute of Mining and Metallurgy, Parkville, pp 181-189

Heithersay PS, Walshe JL (1995) Endeavour 26 North: a porphyry copper-gold deposit in the Late Ordovician shoshonitic Goonumbla volcanic complex, New South Wales, Australia. Econ Geol 90: 1506-1532

Heithersay PS, O'Neill WJ, van der Helder P, Moore CR, Harbon PG (1990) Goonumbla porphyry copper district — Endeavour 26 North, Endeavour 22 and Endeavour 27 copper-gold deposits. In: Hughes FE (ed) Geology of the mineral resources of Australia and Papua New Guinea. The Australasian Institute of Mining and Metallurgy, Parkville, pp 1385-1398 (Australas Inst Min Metall Monogr 14)

Hellman PL, Smith RE, Henderson P (1979) The mobility of the rare earth elements: evidence and implications from selected terrains affected by burial metamorphism. Contrib Mineral Petrol 71: 23-44

Herzig PM, Hannington MD (1995) Hydrothermal activity, vent fauna, and submarine gold mineralization at alkaline fore-arc seamounts near Lihir Island, P.N.G. In: Pacific Rim Congress 95, 19-22 November 1995, Auckland, New Zealand, Proceedings. Australasian Institute for Mining and Metallurgy, Carlton South, pp 279-284

Hickson RJ (1991) Grasberg open-pit: Ertsberg's big brother comes onstream in Indonesia. Mining Engng 43 (3): 385-391

Hildreth W, Drake RE (1992) Volcan Quizapu, Chilean Andes. Bull Volcanol 54: 93-125

Hill KC, Gleadow AJW (1989) Uplift and thermal history of the Papuan Fold Belt, Papua New Guinea: apatite fission track analysis. Austr J Earth Sci 36: 515-539

Hofmann AW, Jochum KP, Seufert PM, White WM (1986) Nb and Pb in oceanic basalts: new constraints on mantle evolution. Earth Planet Sci Lett 79: 33-45

Hoke L (1990) The Altkristallin of the Kreuzeck Mountains, SE Tauern Window, Eastern Alps; basement crust in a convergent plate boundary zone. Jb Geol Bundesanstalt Wien 133: 5-87

Hole MJ, Saunders AD, Marriner GF, Tarney J (1984) Subduction of pelagic sediments: implications for the origin of Ce-anomalous basalts from the Mariana Islands. J Geol Soc 141: 453-472

Höll R, Maucher A (1968) Genese und alter der scheelit-magnesit-lagerstätte tux. Sitzber Bayer Akad Wiss, Math Naturw (München) Kl 1967: 1-11

Holland HD (1972) Granites, solutions and base metals. Econ Geol 67: 281-301

Hollister VF (1975) An appraisal of the nature and source of porphyry copper deposits. Mineral Sci Engng 7: 225-233

Holm PM, Lou S, Nielsen A (1982) The geochemistry and petrogenesis of the lavas of the Vulsinian district, Roman Province, Central Italy. Contrib Mineral Petrol 80: 367-378

Holmes A (1950) Petrogenesis of katungite and its associates. Am Mineral 35: 772-792

Hoogvliet H (1993) The Ladolam gold deposit — at least two major events. Austr J Mining September 1993: 20

Hooper B, Heithersay PS, Mills MB, Lindhorst JW, Freyberg J (1996) Shoshonite-hosted Endeavour 48 porphyry copper-gold deposit, Northparkes, central New South Wales. Austr J Earth Sci 43: 279-288

Horn CM, Pain AM, Newton AW (1989) Mineral resources of the Kapunda district council area. S Austr Dept Mines Energy Rep 89/44

Horton DJ (1978) Porphyry-type copper-molybdenum mineralization belts in Eastern Queensland, Australia. Econ Geol 73: 904-921

Huijsmans JPP, Barton M, Salters VJM (1988) Geochemistry and evolution of the calc-alkaline volcanic complex of Santorini, Aegean Sea, Greece. J Volcanol Geotherm Res 34: 283-306

Hutchison CS, Jezek PA (1978) Banda Arc of eastern Indonesia: petrography, mineralogy and chemistry of the volcanic rocks. In: Proceedings of third regional conference on the geology and mineral resources of Southeast Asia. Smithsonian Institute, Department of Mineral Science, Washington DC, pp 607-619

Iddings JP (1895) Absarokite-shoshonite-banakite series. J Geol 3: 935-959

Ionov DA, Hofmann AW (1995) Nb-Ta-rich mantle amphiboles and micas: implications for subduction-related metasomatic trace element fractionations. Earth Planet Sci Lett 131: 341-356

Introcaso A, Liou A, Ramos VA (1987) La estructura profunda de las Sierras de Cordoba. Revista Asoc Geol Argent 42: 177-187

Jakes P, Smith IE (1970) High-potassium calc-alkaline rocks from Cape Nelson, eastern Papua. Contrib Mineral Petrol 28: 259-271

Jannas RR, Beane RE, Ahler BA, Brosnahan DR (1990) Gold and copper mineralization at the El Indio deposit, Chile. In: Hedenquist JW, White NC, Siddeley G (eds), Epithermal gold mineralization of the Circum-Pacific; geology, geochemistry, origin and exploration, II. J Geochem Expl 36: 233-266

Jaques AL (1976) High-K_2O island-arc volcanic rocks from the Finisterre and Adelbert Ranges, northern Papua New Guinea. Bull Geol Soc Am 87: 861-867

Jaques AL, Lewis JD, Smith CB (1986) The kimberlites and lamproites of Western Australia. Geol Surv West Austr Bull 132

Jensen LS (1978) Larder Lake synoptic mapping project, districts of Cochrane and Timiskaming. Ontario Geol Surv Misc Paper 94: 64-69

Jiang S-Y, Palmer MR, Xue C-J, Li Y-H (1994) Halogen-rich scapolite-biotite rocks from the Tongmugou Pb-Zn deposit, Qinling, northwestern China: implications for the ore-forming processes. Mineral Mag 58: 543-552

Jones GJ (1985) The Goonumbla porphyry copper deposits, N.S.W. Econ Geol 80: 591-613

Joplin GA (1968) The shoshonite association — a review. J Geol Soc Austr 15: 275-294

Joplin GA, Kiss E, Ware NG, Widdowson JR (1972) Some chemical data on members of the shoshonite association. Mineral Mag 38: 936-945

Jordan TE, Allmendinger RW (1986) The Sierras Pampeanas of Argentina: a modern analogue of Rocky Mountain foreland deformation. Am J Svi 286: 737-764

Jordan TE, Douglas RC (1980) Paleogeography and structural development of the Late Pennsylvanian to Early Permian Oquirrh Basin, northwestern Utah. In: Fouch TD, Mahathan ER (eds) Paleozoic paleogeography of the west-central United States; Rocky Mountain paleogeography symposium 1. Society of Economic Paleontologists and Mineralogists, Rocky Mountain Section, Denver, pp 217-230

Kavalieris I, Gonzalez JM (1990) Geology and mineralization of the Pao Prospect: a high sulphidation epithermal gold system hosted in alkaline volcanics, North Luzon, Philippines. In: Pacific Rim Congress 90, 6-12 May 1990, Gold Coast, Queensland, Proceedings. Australasian Institute of Mining and Metallurgy, Parkville, pp 413-418

Kavalieris I, Van Leeuwen TM, Wilson M (1992) Geological setting and styles of mineralization, north arm of Sulawesi, Indonesia. J Southeast Asian Earth Sci 7: 113-129

Kay SM, Abbruzzi JM (1996) Magmatic evidence for Neogene lithospheric evolution of the central Andean "flat-slab" between 30°S and 32°S. Tectonophysics 259: 15-28

Kay RW, Gast PW (1973) The rare earth content and origin of alkali-rich basalts. J Geol 81: 653-682

Kay SM, Gordillo CE (1994) Pocho volcanic rocks and the melting of depleted continental lithosphere above a shallowly dipping, subduction zone in the Andes. Contrib Mineral Petrol 117: 25-44

Kay SM, Maksaev V, Moscoso R, Mpodozis C, Nasi C (1987) Probing the evolving Andean lithosphere: mid-late Tertiary magmatism in Chile (29°-30°30'S) over the modern zone of subhorizontal subduction. J Geophys Res 92: 6173-6189

Kay SM, Maksaev V, Moscoso R, Mpodozis C, Nasi C, Gordillo CE (1988) Tertiary Andean magmatism in Chile and Argentina between 28° S and 33° S: correlation of magmatic chemistry with a changing Benioff zone. J S Amer Earth Sci 1: 21-38

Keays RR (1982) Palladium and iridium in komatiites and associated rocks: application to petrogenetic problems. In: Arndt NT, Nisbet EG (eds) Komatiites. Allen and Unwin, London, pp 435-457

Keith JD, Whitney JA, Cannan TM, Hook C, Hattori K (1995) The role of magmatic sulphides and mafic alkaline magmatism in the formation of giant porphyry and vein systems: examples from the Bingham and Tintic mining districts, Utah. In: Clark AH (ed) Giant Ore Deposits - II. QminEx, Department of Geological Sciences, Queens University, Kingston, pp 316-339

Keith JD, Whitney JA, Hattori K, Ballantyne GH, Christiansen EH, Barr DL, Cannan TM, Hook CJ (1997) The role of magmatic sulfides and mafic alkaline magmas in the Bingham and Tintic Mining Districts, Utah. J Petrol 38: 1679-1690

Kelemen PB, Johnson KTM, Kinzler RJ, Irving AJ (1990) High field strength element depletions in arc basalts due to mantle-magma interactions. Nature 345: 521-524

Keller J (1974) Petrology of some volcanic rock series of the Aeolian Arc, Southern Tyrrhenian Sea: calc-alkaline and shoshonitic associations. Contrib Mineral Petrol 46: 29-47

Kelly JL (1977) Geology of the Twin Buttes copper deposit, Pima County, Arizona. SME-AIME Trans 262: 110-116

Kelly KD, Romberger SB, Beaty DW, Pontius JA, Snee LW, Stein HJ, Thompson TB (1998) Geochemical and geochronological constraints on the genesis of Au-Te deposits at Cripple Creek, Colorado. Econ Geol 93: 981-1012

Kesler SE (1997) Metallogenic evolution of convergent margins: selected ore deposit models. Ore Geol Rev 12: 153-171

Kesler SE, Issigonis MJ, Brownlow AH, Damon PE, Moore WJ, Northcote KE, Preto VA (1975) Geochemistry of biotites from mineralized and barren intrusive systems. Econ Geol 70: 559-567

Keyser F (1961) Misima Island — geology and gold mineralisation. Bur Mineral Resources, Austr, BMR Rep 57: 1-36

Kilinc IA, Burnham CW (1972) Partitioning of chloride between a silicate melt and coexisting aqueous phase from 2 to 8 kilobars. Econ Geol 67: 231-235

Kirchner JG (1979) Petrographic significance of a carbonate-rich lamprophyre from Squaw Creek, northern Black Hills, South Dakota. Am Mineral 64: 986-992

Kirkham RV, Margolis J (1995) Overview of the Sulphurets area, NW-British Columbia. In: Schroeter TG (ed) Porphyry deposits of the NW-Cordillera of North America. Canadian Institute of Mining, Metallurgy and Petroleum, Montreal, pp 473-483 (Can Inst Min Metall Petrol Spec Vol 46)

Kirkham RV, Sinclair WD (1996) Porphyry copper, gold, molybdenum, tungsten, tin, silver. In: Eckstrand OR, Sinclair WD, Thorpe RI (eds) Geology of Canadian mineral deposit types. Geological Survey of Canada, Ottawa, pp 405-430 (Geol Canada 8)

Knittel U, Burton CK (1985) Polillo Island (Philippines): molybdenum mineralization in an island arc. Econ Geol 80: 2013-2018

Kontak DJ, Clark AH, Farrar E, Pearce TH, Strong DF, Baadsgaard H (1986) Petrogenesis of a Neogene shoshonite suite, Cerro Moromoroni, Puno, southeastern Peru. Can Mineral 24: 117-135

Kullerud K (1995) Chlorine, titanium and barium-rich biotites: factors controlling biotite composition and the implications for garnet-biotite geothermometry. Contrib Mineral Petrol 120: 42-59

Kwak TAP (1990) Geochemical and temperature controls on ore mineralization at the Emperor gold mine, Vatukoula, Fiji. In: Hedenquist JW, White NC, Siddeley G (eds), Epithermal gold mineralization of the Circum-Pacific; geology, geochemistry, origin and exploration, II. J Geochem Expl 36: 297-337

Lahusen L (1972) Schicht- und zeitgebundene Antimonit-Scheelit-vorkommen und Zinnobervererzung in Kärnten Osttirol/Österreich. Mineralium Deposita 7: 31-60

Lange RA, Carmichael ISE (1990) Hydrous basaltic andesites associated with minette and related lavas in Western Mexico. J Petrol 31: 1225-1259

Lanier G, John EC, Swensen AJ, Reid J, Bard CE, Caddey SW, Wilson JC (1978a) General geology of the Bingham mine, Bingham Canyon, Utah. Econ Geol 73: 1228-1241

Lanier G, Raab WJ, Folsom RB, Cone S (1978b) Alteration of equigranular monzonite, Bingham mining district, Utah. Econ Geol 73: 1270-1286

Large RR, Huston DL, McGoldrick PJ, Ruxton PA (1989) Gold distribution and genesis in Australian volcanogenic massive sulphide deposits and their significance for gold transport models. In: Keays RR, Ramsay WRH, Groves DI (eds), The geology of gold deposits: the perspective in 1988. The Economic Geology Publishing Company, El Paso, pp 520-536 (Econ Geol Monogr 6)

Laubscher H (1988) Material balance in Alpine orogeny. Bull Geol Soc Am 100: 1313-1328

Leat PT, Thompson RN, Morrison MA, Hendry GL, Dickin AP (1988) Silicic magma derived by fractional-crystallization from Miocene minette, Elkhead Mountain, Colorado. Mineral Mag 52: 577-586

Le Maitre RW (1962) Petrology of volcanic rocks, Gough Island, South Atlantic. Bull Geol Soc Am 73: 1309-1340

Le Maitre RW (1982) Numerical petrology: statistical interpretation of geochemical data. Elsevier, Amsterdam, 291 pp (Elsevier Developments in Petrology 8)

Le Maitre RW, ed (1989) A classification of igneous rocks and glossary of terms: recommendations of the International Union of Geological Sciences Subcommission on the Systematics of Igneous Rocks. Blackwell Scientific Publications, Oxford, 193 pp

Le Roex A (1985) Geochemistry, mineralogy and magmatic evolution of the basaltic and trachytic lavas from Gough Island, South Atlantic. J Petrol 26: 149-186

Leterrier J, Yumono YS, Soeria-Atmadja R, Maury RC (1990) Potassic volcanism in Central Java and South Sulawesi, Indonesia. J Southeast Asian Earth Sci 4: 171-181

Levi B, Nyström JO, Thiele R, Aberg G (1988) Geochemical trends in Mesozoic-Tertiary volcanic rocks from the Andes in central Chile and its tectonic implications. J South American Earth Sci 1: 63-74

Lewis JD (1987) The geology and geochemistry of the lamproites of the Ellendale Field, West Kimberley Region, Western Australia. MSc Thesis, The University of Western Australia, Perth

Lewis RW, Wilson GI (1990) Misima gold deposit. In: Hughes FE (ed) Geology of the mineral resources of Australia and Papua New Guinea. The Australasian Institute of Mining and Metallurgy, Parkville, pp 1741-1745 (Australas Inst Min Metall Monogr 14)

Lin PN, Stern RJ, Bloomer SH (1989) Shoshonitic volcanism in the northern Marianas arc: 2. LILE and REE abundances: evidence for the source of incompatible element enrichments in intra oceanic arcs. J Geophys Res 94: 4497-4514

Lindgren W (1933) Mineral deposits, 4th edition. McGraw-Hill Book Company, New York: 797 pp

Llambias EG (1972) Estructura del grup volcanico Farallon Negro, Catamarca, Rupublica Argentina. Revista Asoc Geol Argent 27: 161-169

Loferski PJ, Ayuso RA (1995) Petrography and mineral chemistry of the composite Deboullie pluton, northern Maine, U.S.A.: implications for the genesis of Cu-Mo mineralization. Chem Geol 123: 89-105

Lowenstern JB (1994) Dissolved volatile concentrations in an ore-forming magma. Geology 22: 893-896

Luhr JF, Kyser TK (1989) Primary igneous analcime: the Colima minettes. Am Mineral 74: 216-223

Luhr JF, Allan JF, Carmichael ISE, Nelson SA, Hasenaka T (1989) Primitive calc-alkaline and alkaline rock types from the western Mexican volcanic belt. J Geophys Res 94: 4515-4530

MacDonald GD, Arnold LC (1994) Geological and geochemical zoning of the Grasberg igneous complex, Irian Jaya, Indonesia. J Geochem Expl 50: 143-178

Magenheim AJ, Spivack AJ, Michael PJ, Gieskes JM (1995) Chlorine stable isotope composition of the oceanic crust: implications for Earth's distribution of chlorine. Earth Planet Sci Lett 131: 427-432

Marcelot G, Dupuy C, Girod M, Maury RC (1983) Petrology of Futuna Island lavas (New Hebrides): an example of calc-alkaline magmatism associated with the initial stages of back-arc spreading. Chem Geol 38: 23-37

McDonald R, Rock NMS, Rundle CC, Russell OJ (1986) Relationships between late Caledonian lamprophyric and acidic magmas in a differentiated dyke, SW Scotland. Mineral Mag 50: 547-557

McDonald R, Upton BJ, Collerson KD, Hearn BC, James D (1992) Potassic mafic lavas of the Bearpaw Mountains, Montana: mineralogy, chemistry and origin. J Petrol 33: 305-346

McKenzie DE, Chappell BW (1972) Shoshonitic and calc-alkaline lavas from the highlands of Papua New Guinea. Contrib Mineral Petrol 35: 50-62

McLennan SM, Taylor SR (1981) Role of subducted sediments in island arc magmatism: constraints from REE patterns. Earth Planet Sci Lett 54: 423-430

Meen JK (1987) Formation of shoshonites from calc-alkaline basalt magmas: geochemical and experimental constraints from the type locality. Contrib Mineral Petrol 97: 333-351

Mehnert HH, Lipman PW, Steven TA (1973) Age of mineralization at Summitville, Colorado, as indicated by K-Ar dating of alunite. Econ Geol 68: 399-401

Meijer A, Reagan M (1981) Petrology and geochemistry of the island of Sarigan in the Mariana arc; calc-alkaline volcanism in an oceanic setting. Contrib Mineral Petrol 77: 337-354

Menzies MA, Hawkesworth CJ, eds (1987) Mantle metasomatism. Academic Press, London, 472 pp

Metrich N, Rutherford MJ (1992) Experimental study of chlorine behavior in hydrous silicic melts. Geochim Cosmochim Acta 56: 607-616

Mitchell AHG, Garson MS (1981) Mineral deposits and global tectonic setting. Academic Press, London, 421 pp

Mitchell AHG, McKerrow WS (1975) Analogous evolution of the Burma orogen and the Scottish Caledonides. Bull Geol Soc Am 86: 305-315

Mitchell AHG, Warden AJ (1971) Geological evolution of the New Hebrides island arc. J Geol Soc 127: 501-529

Mitchell RH (1986) Kimberlites: mineralogy, geochemistry, and petrology. Plenum Press, New York, 442 pp

Mitchell RH (1989) Aspects of the petrology of kimberlites and lamproites: some definitions and distinctions. In: Ross J, Jaques AL, Ferguson J, Green DH, O'Reilly SY, Danchin RV, Janse AJA (eds) Kimberlites and related rocks. Geological Society of Australia, Sydney, pp 7-45 (Geol Soc Austr Spec Publ 14)

Mitchell RH, Bell K (1976) Rare earth element geochemistry of potassic lavas from the Birunga and Toro-Ankole regions of Uganda, Africa. Contrib Mineral Petrol 58: 293-303

Mitchell RH, Bergman SC (1991) Petrology of lamproites. Plenum Press, New York, 447 pp

Mitchell RH, Keays RR (1981) Abundance and distribution of gold, palladium and iridium in some spinel and garnet lherzolites: implications for the nature and origin of precious metal-rich intergranular components in the upper mantle. Geochim Cosmochim Acta 45: 2425-2442

Mittempergher M (1965) Volcanism and petrogenesis in the San Venanzo area, Italy. Bull Volcanol 28: 1-12

Mogessie A, Tessadri R, Veltman CB (1990) EMP-AMPH — A hypercard program to determine the name of an amphibole from electron microprobe analysis according to the international mineralogical association scheme. Comput Geosci 16: 309-330

Moore G, Righter K, Carmichael ISE (1995) The effect of dissolved water on the oxidation state of iron in natural silicate liquids. Contrib Mineral Petrol 120: 170-179

Morris BJ (1990) Kanmantoo trough geological investigations, Karinya syncline: Truro lamprophyres. S Austr Dept Mines Energy Report 91/29

Morrison GW (1980) Characteristics and tectonic setting of the shoshonite rock association. Lithos 13: 97-108

Morrison MA (1978) The use of "immobile" trace elements to distinguish the palaeotectonic affinities of metabasalts: applications to the Palaeocene basalts of Mull and Skye, Northwest Scotland. Earth Planet Sci Lett 39: 407-416

Mortensen JK (1993) U-Pb geochronology of the eastern Abitibi Subprovince. Part 2: Noranda - Kirkland Lake area. Can J Earth Sci 30: 29-41

Moyle AJ, Doyle BJ, Hoogvliet H, Ware AR (1990) Ladolam gold deposit, Lihir Island. In: Hughes FE (ed) Geology of the mineral resources of Australia and Papua New Guinea. The Australasian Institute of Mining and Metallurgy, Parkville, pp 1793-1805 (Australas Inst Min Metall Monogr 14)

Muenow DW, Garcia MO, Aggrey KE, Bednarz U, Schmincke HU (1990) Volatiles in submarine glasses as a discriminant of tectonic origin: application to the Troodos ophiolite. Nature 343: 159-161

Müller D (1993) Shoshonites and potassic igneous rocks: indicators for tectonic setting and mineralization potential of modern and ancient terranes. PhD Thesis, The University of Western Australia, Perth

Müller D, Forrestal P (1998) The shoshonite porphyry Cu-Au association at Bajo de la Alumbrera, Catamarca Province, Argentina. Mineral Petrol 64: 47-64

Müller D, Groves DI (1993) Direct and indirect associations between potassic igneous rocks, shoshonites and gold-copper deposits. Ore Geol Rev 8: 383-406

Müller D, Stumpfl EF, Taylor WR (1992a) Shoshonitic and alkaline lamprophyres with elevated Au and PGE concentrations from the Kreuzeck Mountains, Eastern Alps, Austria. Mineral Petrol 46: 23-42

Müller D, Rock NMS, Groves DI (1992b) Geochemical discrimination between shoshonitic and potassic volcanic rocks from different tectonic settings: a pilot study. Mineral Petrol 46: 259-289

Müller D, Morris BJ, Farrand MG (1993a) Potassic alkaline lamprophyres with affinities to lamproites from the Karinya Syncline, South Australia. Lithos 30: 123-137

Müller D, Groves DI, Stumpfl EF (1993b) Potassic igneous rocks and shoshonites as potential exploration targets. In: IAVCEI, General Assembly, Canberra, September 1993, Ancient Volcanism and Modern Analogues, Abstracts. International Association for Volcanology and Chemistry of the Earth's Interior, Canberra, p 76

Müller D, Heithersay PS, Groves DI (1994) The shoshonite-porphyry Cu-Au association in the Goonumbla district, N.S.W., Australia. Mineral Petrol 51: 299-321

Munoz JL (1984) F-OH and Cl-OH exchange in micas with applications to hydrothermal ore deposits. In: Bailey SW (ed) Micas. American Mineralogical Society, Michigan, pp 469-493 (Rev Mineral 13)

Mutschler FE, Mooney TC (1993) Precious-metal deposits related to alkalic igneous rocks: provisional classification, grade-tonnage data and exploration frontiers. In: Kirkham RV, Sinclair WD, Thorpe RI, Duke JM (eds) Mineral deposit modelling. Geological Association of Canada, Toronto, pp 479-520 (Geol Assoc Canada Spec Pap 40)

Mutschler FE, Griffin ME, Scott-Stevens D, Shannon SS (1985) Precious metal deposits related to alkaline rocks in the North American Cordillera — an interpretative view. Trans Geol Soc S Afr 88: 355-377

Myers JS (1993) Precambrian history of the Western Australian Craton and adjacent orogens. Ann Rev Earth Planet Sci 21: 453-485

Naumov VB, Kovalenko VI, Dorofeeva VA (1998) Fluorine concentration in magmatic melts: evidence from inclusions in minerals. Geochem Int 36: 117-127

Needham RS, De Ross GJ (1990) Pine Creek Inlier — regional geology and mineralization. In: Hughes FE (ed) Geology of the mineral resources of Australia and Papua New Guinea. The Australasian Institute of Mining and Metallurgy, Parkville, pp 727-737 (Australas Inst Min Metall Monogr 14)

Needham RS, Roarty MJ (1980) An overview of metallic mineralization in the Pine Creek Geosyncline. In: Ferguson J, Goleby AB (eds) Proceedings of international uranium symposium on the Pine Creek Geosyncline, Sydney, 1979. International Atomic Energy Agency, Vienna, pp 157-174

Needham RS, Stuart-Smith PG, Page RW (1988) Tectonic evolution of the Pine Creek Inlier, Northern Territory. Precambr Res 41: 543-564

Nelson DR, McCulloch MT, Sun SS (1986) The origins of ultrapotassic rocks as inferred from Sr, Nd and Pb isotopes. Geochim Cosmochim Acta 50: 231-245

Nicholls J (1969) Studies of the volcanic petrology of the Navajo-Hopi area, Arizona. PhD Thesis, University of California, Berkeley

Nicholls J, Carmichael ISE (1969) A commentary on the absarokite-shoshonite-banakite series of Wyoming, USA. Schweiz Mineral Petrol Mitt 49: 47-64

Nicholson PM, Eupene GS (1990) Gold deposits of the Pine Creek Inlier. In: Hughes FE (ed) Geology of the mineral resources of Australia and Papua New Guinea. The Australasian Institute of Mining and Metallurgy, Parkville, pp 739-742 (Australas Inst Min Metall Monogr 14)

Ninkovich D, Hays JD (1972) Mediterranean island arcs and origin of high potash volcanoes. Earth Planet Sci Lett 16: 331-345

O'Driscoll EST (1983) Deep tectonic foundations of the Eromanga Basin. Austr Petrol Explor Ass J 23: 5-17

Oxburgh E (1966) Superimposed fold system in the Altkristallin rocks on the southeast margin of the Tauern Window. Verh Geol Bundesanstalt Wien Austria 1966: 33-46

Page RW (1988) Geochronology of Early to Middle Proterozoic fold belts in northern Australia: a review. Precambr Res 41: 1-19

Parry WT, Ballantyne GH, Wilson JC (1978) Chemistry of biotites and apatites from a vesicular quartz latite porphyry plug at Bingham, Utah. Econ Geol 73: 1308-1314

Paul DK, Crocket JH, Nixon PH (1979) Abundance of palladium, iridium and gold in kimberlites and associated nodules. In: Boyd FR, Meyer HAO (eds) Kimberlites, diatremes and diamonds. Proceedings of 2nd international kimberlite conference, volume 1. American Geophysical Union, Washington, pp 272-279

Peach CL, Mathez EA, Keays RR, Reeves SJ (1994) Experimentally determined sulphide melt-silicate melt partition coefficients for iridium and palladium. Chem Geol 117: 361-377

Pearce JA (1976) Statistical analysis of major element patterns in basalt. J Petrol 17: 15-43

Pearce JA (1982) Trace element characteristics of lavas from destructive plate boundaries. In: Thorpe RS (ed) Andesites. Wiley, New York, pp 525-548

Pearce JA (1983) Role of sub-continental lithosphere in magma genesis at active continental margins. In: Hawkesworth CJ, Norry MJ (eds) Continental basalts and mantle xenoliths. Shiva, Nantwich, pp 230-249

Pearce JA (1987) An expert system for the tectonic characterization of ancient volcanic rocks. J Volcanol Geotherm Res 32: 51-65

Pearce JA, Cann JR (1973) Tectonic setting of basic volcanic rocks determined using trace element analyses. Earth Planet Sci Lett (19: 290-300

Pearce JA, Norry MJ (1979) Petrogenetic implications of Ti, Zr, Y and Nb variations in volcanic rocks. Contrib Mineral Petrol 69: 33-47

Pearce JA, Harris NBW, Tindle AG (1984) Tectonic interpretation of granitic rocks. J Petrol 25: 956-983

Peccerillo A (1992) Potassic and ultrapotassic rocks: compositional characteristics, petrogenesis, and geologic significance. IUGS Episodes 15: 243-251

Peccerillo A, Taylor SR (1976a) Geochemistry of Eocene calc-alkaline volcanic rocks from the Kastomonon area, northern Turkey. Contrib Mineral Petrol 58: 63-81

Peccerillo A, Taylor SR (1976b) Rare earth elements in east Carpathian volcanic rocks. Earth Planet Sci Lett 32: 121-126

Pe-Piper G (1980) Geochemistry of Miocene shoshonites, Lesbos, Greece. Contrib Mineral Petrol 72: 387-396

Perkins C, McDougall I, Claoué-Long J, Heithersay PS (1990a) ^{40}Ar/^{39}Ar and U-Pb geochronology of the Goonumbla porphyry Cu-Au deposits, N.S.W, Australia. Econ Geol 85: 1808-1824

Perkins C, McDougall I, Claoué-Long J (1990b) Dating of ore deposits with high precision: examples from the Lachlan Fold Belt, N.S.W, Australia. In: Proceedings of Pacific rim congress 90. The Australasian Institute of Mining and Metallurgy, Parkville, pp 105-112

Perkins C, McDougall I, Walshe JL (1992) Timing of shoshonitic magmatism and gold mineralization, Sheahan-Grants and Glendale, New South Wales. Austr J Earth Sci 39: 99-110

Perring CS (1988) Petrogenesis of the lamprophyre-"porphyry" suite from Kambalda, Western Australia. In: Ho SE, Groves DI (eds) Advances in understanding Precambrian gold deposits volume II. Geology Department & University Extension, The University of Western Australia, Perth, pp 277-294 (Geol Dept & Univ Extension, Univ West Austr, Publ 12)

Perring CS, Barley ME, Cassidy KF, Groves DI, McNaughton NJ, Rock NMS, Bettenay LF, Golding SD, Hallberg JA (1989a) The association of linear orogenic belts, mantle-crustal magmatism and Archean gold mineralization in the Eastern Yilgarn Block of Western Australia. In: Keays RR, Ramsay WRH, Groves DI (eds), The geology of gold deposits: the perspective in 1988. The Economic Geology Publishing Company, El Paso, pp 571-584 (Econ Geol Monogr 6)

Perring CS, Rock NMS, Golding SD, Roberts DE (1989b) Criteria for the recognition of metamorphosed or altered lamprophyres: a case study from the Archean of Kambalda, Western Australia. Precambr Res 43: 215-237

Pitcher WS (1983) Granite type and tectonic environment. In: Hsu K (ed) Mountain building process. Academic Press, London, pp 19-40

Plimer IR, Andrew AS, Jenkins R, Lottermoser BG (1988) The geology and geochemistry of the Lihir gold deposit, Papua New Guinea. Geol Soc Austr Abst 22: 139-143

Poli G, Frey FA, Ferrara G (1984) Geochemical characteristics of the South Tuscany (Italy) volcanic province: constraints on lava petrogenesis. Chem Geol 43: 203-221

Powell CMcA (1984) Ordovician to earliest Silurian: marginal sea and island arc. In: Veevers JJ (ed) Phanerozoic Earth history of Australia. Clarendon Press, Oxford, pp 290-312

Preiss WV (1987) The Adelaide Geosyncline — late Proterozoic stratigraphy, sedimentation, paleontology and tectonics. Bull Geol Surv S Aust 53: 438 pp

Prider RT (1960) The leucite lamproites of the Fitzroy Basin, Western Australia. J Geol Soc Austr 6: 71-118

Puntodewo SSO, McCaffrey R, Calais E, Bock Y, Rais J, Subarya C, Poewariardi R, Stevens C, Genrich J, Fauzi, Zwick P, Wdowinski S (1994) GPS measurements of crustal deformation within the Pacific-Australia plate boundary zone in Irian Jaya, Indonesia. Tectonophysics 237: 141-153

Reagan MK, Gill JB (1989) Coexisting calc-alkaline and high-Nb-basalts from Turrialba volcano, Costa Rica: implications for residual titanates in arc magma sources. J Geophys Res 94: 4619-4633

Reimann C, Stumpfl EF (1981) Geochemical setting of strata-bound stibnite mineralization in the Kreuzeck Mountains, Austria. Trans Instn Min Metall 90: 126-132

Reimann C, Stumpfl EF (1985) Paleozoic amphibolites, Kreuzeck Mountains, Austria: geochemical variations in the vicinity of mineralization. Mineralium Deposita 20: 69-75

Reyes M (1991) The Andacollo strata-bound gold deposit, Chile, and its position in a porphyry copper-gold system. Econ Geol 86: 1301-1316

Rice CM, Harmon RS, Shepherd TJ (1985) Central City, Colorado: the upper part of an alkaline porphyry molybdenum system. Econ Geol 80: 1769-1796

Richards JP (1990a) The Porgera gold deposit, Papua New Guinea: geology, geochemistry and geochronology. PhD Thesis, The Australian National University, Canberra

Richards JP (1990b) Petrology and geochemistry of alkalic intrusives at the Porgera gold deposit, Papua New Guinea. J Geochem Expl 35: 141-199

Richards JP (1992) Magmatic-epithermal transitions in alkalic systems: Porgera gold deposit, Papua New Guinea. Geology 20: 547-550

Richards JP (1995) Alkalic-type epithermal gold deposits — a review. In: Thompson JFH (ed) Magmas, fluids, and ore deposits. Mineralogical Association of Canada, Toronto, pp 367-400 (Min Assoc Canada Short Course Hbook 23)

Richards JP, Kerrich R (1993) The Porgera gold mine, Papua New Guinea: magmatic hydrothermal to epithermal evolution of an alkalic-type precious metal deposit. Econ Geol 88: 1017-1052

Richards JP, Ledlie I (1993) Alkalic intrusive rocks associated with the Mount Kare gold deposit, Papua New Guinea: comparison with the Porgera intrusive complex. Econ Geol 88: 755-781

Richards JP, Chappell BW, McCulloch MT (1990) Intraplate-type magmatism in a continent-island-arc collision zone: Porgera intrusive complex, Papua New Guinea. Geology 18: 958-961

Richards JP, McCulloch MT, Chappell BW, Kerrich R (1991) Sources of metals in the Porgera gold deposit, Papua New Guinea: evidence from alteration, isotope, and noble metal geochemistry. Geochim Cosmochim Acta 55: 565-580

Ringwood AE (1990) Slab-mantle interactions, 3: petrogenesis of intraplate magmas and structure of the upper mantle. Chem Geol 82: 187-207

Rittmann A (1933) Die geologisch bedingte evolution und differentiation des Somma-Vesuv Magmas. Zeitsch für Vulkanol 15: 8-94

Rock NMS (1977) The nature and origin of lamprophyres: some definitions, distinctions and derivations. Earth Sci Rev 13: 123-169

Rock NMS (1987) The nature and origin of lamprophyres: an overview. In: Fitton JG, Upton BGJ (eds) Alkaline igneous rocks. Geological Society, London, pp 191-226 (Geol Soc Lond Spec Publ 30)

Rock NMS (1988) Numerical geology. Springer-Verlag, Heidelberg, 427 pp (Lecture Notes in Earth Sciences 18)

Rock NMS (1991) Lamprophyres. Blackie, Glasgow, 285 pp

Rock NMS, Finlayson EJ (1990) Petrological affinities of intrusive rocks associated with the giant mesothermal gold deposit at Porgera, Papua New Guinea. J S-E Asian Earth Sci 4: 247-257

Rock NMS, Groves DI (1988a) Do lamprophyres carry gold as well as diamonds? Nature 332: 253-255

Rock NMS, Groves DI (1988b) Can lamprophyres resolve the genetic controversy over mesothermal gold deposits? Geology 16: 538-541

Rock NMS, Syah HH, Davis AE, Hutchison D, Styles MT, Rahayu L (1982) Permian to Recent volcanism in Northern Sumatra, Indonesia: a preliminary study of its distribution, chemistry and peculiarities. Bull Volcanol 45: 127-152

Rock NMS, Duller P, Haszeldine S, Groves DI (1987) Lamprophyres as potential gold exploration targets: some preliminary observations and speculations. In: Ho SE, Groves DI (eds) Recent advances in understanding Precambrian gold deposits. Geology Department & University Extension, The University of Western Australia, Perth, pp 271-286 (Geol Dept & Univ Extension, Univ West Austr, Publ 11)

Rock NMS, Groves DI, Ramsay RR (1988a) Lamprophyres: a girl's best friend? In: Ho SE, Groves DI (eds) Advances in understanding Precambrian gold deposits volume II. Geology Department & University Extension, The University of Western Australia, Perth, pp 295-308 (Geol Dept & Univ Extension, Univ West Austr, Publ 12)

Rock NMS, Hallberg JA, Groves DI, Mather PJ (1988b) Archean lamprophyres in the goldfields of the Yilgarn Block, Western Australia: new indications of their widespread distribution and significance. In: Ho SE, Groves DI (eds) Advances in understanding Precambrian gold deposits volume II. Geology Department & University Extension, The University of Western Australia, Perth, pp 245-275 (Geol Dept & Univ Extension, Univ West Austr, Publ 12)

Rock NMS, Groves DI, Perring CS, Golding SD (1989) Gold, lamprophyres, and porphyries: what does their association mean? In: Keays RR, Ramsay WRH, Groves DI (eds), The geology of gold deposits: the perspective in 1988. The Economic Geology Publishing Company, El Paso, pp 609-625 (Econ Geol Monogr 6)

Rock NMS, Taylor WR, Perring CS (1990) Lamprophyres — what are lamprophyres? In: Ho SE, Groves DI, Bennett JM (eds) Gold deposits of the Archaean Yilgarn Block, Western Australia: nature, genesis and exploration guides. Geology Key Centre & University Extension, The University of Western Australia, Perth, pp 128-135 (Geol Dept & Univ Extension, Univ West Austr, Publ 20)

Roden MF (1981) Origin of coexisting minette and ultramafic breccia, Navajo volcanic field. Contrib Mineral Petrol 77: 195-206

Roden MF, Smith D (1979) Field geology, chemistry and petrology of Buell Park diatreme, Apache county, Arizona. In: Boyd FR, Meyer HAO (eds) Kimberlites, diatremes and diamonds. Proceedings of 2nd international kimberlite conference, volume 1. American Geophysical Union, Washington, pp 364-381

Roedder E (1984) Fluid inclusions. Min Soc Am, Rev Mineral 12: 644 pp

Roegge JS, Logsdon MJ, Young HS, Barr HB, Borcsik M, Holland HD (1974) Halogens in apatites from the Providencia area, Mexico. Econ Geol 69: 229-240

Rogers NW, Bachinski SW, Henderson P, Parry SJ (1982) Origin of potash-rich basic lamprophyres: trace element data from Arizona minettes. Earth Planet Sci Lett 57: 305-312

Rogers NW, Hawkesworth CJ, Parker RJ, Marsh JS (1985) The geochemistry of potassic lavas from Vulsini, central Italy, and implications for mantle enrichment processes beneath the Roman Region. Contrib Mineral Petrol 90: 244-257

Rogers NW, De Mulder M, Hawkesworth CJ (1992) An enriched mantle source for potassic basanites: evidence from Karisimbi volcano, Virunga volcanic province, Rwanda. Contrib Mineral Petrol 111: 543-556

Rogers NW, James D, Kelley SP, De Mulder M (1998) The generation of potassic lavas from the eastern Virunga Province, Rwanda. J Petrol 39: 1223-1247

Rowins SM, Cameron EM, Lalonde AE, Ernst RE (1993) Petrogenesis of the late Archean syenitic Murdock Creek pluton, Kirkland Lake, Ontario: evidence for an extensional tectonic setting. Can Mineral 31: 219-244

Rush PM, Seegers HJ (1990) Ok Tedi copper-gold deposits. In: Hughes FE (ed) Geology of the mineral deposits of Australia and Papua New Guinea. The Australasian Institute of Mining and Metallurgy, Parkville, pp 1747-1754 (Australas Inst Min Metall Monogr 14)

Russell PJ (1990) Woodlark Island gold deposits. In: Hughes FE (ed) Geology of the mineral resources of Australia and Papua New Guinea. The Australasian Institute of Mining and Metallurgy, Parkville, pp 1735-1739 (Australas Inst Min Metall Monogr 14)

Russell PJ, Finlayson EJ (1987) Volcanic-hosted epithermal mineralization on Woodlark Island, Papua New Guinea. In: Pacific Rim Congress 87, 26-29 August 1987, Gold Coast, Queensland, Proceedings. Australasian Institute of Mining and Metallurgy, Parkville, pp 381-385

Sahama AG (1974) Potassium-rich alkaline rocks. In: Sørensen H (ed) The alkaline rocks. Wiley, New York, pp 96-109

Salek H (1976) The silver occurrence study of a sample from the Twin Buttes project, Pima County, Arizona. Anaconda Company Report, 1976

Salfity JA (1985) Lineamentos transversales al rumbo Andino en al noroeste Argentino. In: Arias FJ, Camano BP, Espinoza RS (eds) IV Congreso Geologica Chileno, 19-24 August 1985, Antofagasta, Chile, Proceedings. Universidad del Norte Chile, Departmento de Geociencias, Antofagasta, pp 119-137

Sato M, Wright TC (1966) Oxygen fugacity directly measured in magmatic gases. Science 153: 1103-1105

Saunders AD, Tarney J, Weaver SD (1980) Transverse geochemical variations across the Antarctic peninsula: implications for the genesis of calc-alkaline magmas. Earth Planet Sci Lett 6: 344-360

Saunders AD, Norry MJ, Tarney J (1991) Fluid influence on the trace element compositions of subduction zone magmas. Trans R Soc London A335: 377-392

Savelli C (1967) The problem of rock assimilation by Somma-Vesuvius magma; I. Composition of Somma and Vesuvius lavas. Contrib Mineral Petrol 16: 328-353

Scheibner E (1972) Tectonic concepts and tectonic mapping. NSW Geol Surv Rec 14: 37-83

Scheibner E (1974) Lachlan fold belt. Definition and review of structural elements. In: Markham NL, Basden H (eds) The mineral deposits of New South Wales. NSW Geological Survey, Sydney, pp 109-113 (NSW Geol Surv Rec 16)

Schreiber U, Schwab K (1991) Geochemistry of Quaternary shoshonitic lavas related to the Calama-Olacapato-El Toro Lineament, NW-Argentina. J S Amer Earth Sci 4: 73-85

Schwab K, Lippolt H (1974) K-Ar mineral ages and late Cenozoic history of the Salar de Canchari area (Argentine Puna). In: Gonzalez FO (ed) Proceedings of the IAVCEI Symposium on Andean and Antarctic volcanology problems, September 1974, Santiago, Chile. International Association of Volcanology and Chemistry of the Earth's Interior, Rome, pp 698-714

Scott-Smith BH, Skinner EMW (1982) A new look at Prairie Creek, Arkansas. Terra Cognita 2: 210

Scott-Smith BH, Skinner EMW (1984) Diamondiferous lamproites. J Geol 92: 433-438

Setterfield TN (1991) Evolution of the Tavua caldera and associated hydrothermal systems, Fiji. PhD Thesis, University of Cambridge, Cambridge

Setterfield TN, Eaton PC, Rose WJ, Sparks RSJ (1991) The Tavua caldera, Fiji: a complex shoshonitic caldera formed by concurrent faulting and downsagging. J Geol Soc 148: 115-127

Setterfield TN, Mussett AE, Oglethorpe RDJ (1992) Magmatism and associated hydrothermal activity during the evolution of the Tavua caldera: ^{40}Ar-^{39}Ar dating on the volcanic, intrusive and hydrothermal events. Econ. Geol 87: 1130-1140

Sheppard S (1992) An early Proterozoic shoshonitic lamprophyre-granite association and its relationship to the Tom's Gully gold deposit, Mt. Bundey, Northern Territory, Australia. PhD Thesis, The University of Western Australia, Perth

Sheppard S (1995) Hybridization of shoshonitic lamprophyre and calc-alkaline granite magma in the Early Proterozoic Mt Bundey igneous suite, Northern Territory. Aust J Earth Sci 42: 173-185

Sheppard S, Taylor WR (1992) Barium- and LREE-rich, olivine-mica lamprophyres with affinities to lamproites, Mt. Bundey, Northern Territory, Australia. Lithos 28: 303-325

Siddeley G, Araneda R (1986) The El Indio - Tambo gold deposits, Chile. In: Macdonald AJ (ed) Proceedings of Gold '86, an international symposium on the geology of gold deposits. Gold '86, Toronto, pp 445-456

Sighinolfi GP, Gorgoni C, Mohamed AH (1984) Comprehensive analysis of precious metals in some geological standards by flameless Atomic Absorption Spectroscopy. Geostandards Newsletter 8: 25-29

Sillitoe RH (1972) A plate tectonic model for the origin of porphyry copper deposits. Econ Geol 67: 184-197

Sillitoe RH (1979) Some thoughts on gold-rich porphyry copper deposits. Mineralium Deposita 14: 161-174

Sillitoe RH (1991) Gold metallogeny of Chile — an introduction. Econ Geol 86: 1187-1205

Sillitoe RH (1993) Giant and bonanza gold deposits in the epithermal environment: assessment of potential genetic factors. In: Whiting BH, Hodgson CJ, Mason R (eds), Giant ore deposits. The Economic Geology Publishing Company, El Paso, pp 125-156 (Econ Geol Spec Publ 2)

Sillitoe RH (1997) Characteristics and controls of the largest porphyry copper-gold and epithermal gold deposits in the circum-Pacific region. Aust J Earth Sci 44: 373-388

Sillitoe RH, Camus F (1991) A special issue devoted to the gold deposits in the Chilean Andes — preface. Econ Geol 86: 1153-1154

Sillitoe RH, Gappe IM (1984) Philippine porphyry copper deposits: geologic setting and characteristics. U.N.-ESCAP, CCOP Tech. Publ 14

Skinner EMW (1989) Contrasting group 2 and group 1 kimberlite petrology: towards a genetic model for kimberlites. In: In: Ross J, Jaques AL, Ferguson J, Green DH, O'Reilly SY, Danchin RV, Janse AJA (eds) Kimberlites and related rocks. Geological Society of Australia, Sydney, pp 528-544 (Geol Soc Austr Spec Publ 14)

Smith CB, Gurney JJ, Skinner EMW, Clement CR, Ebrahim N (1985) Geochemical character of southern African kimberlites: a new approach based upon isotopic constraints. Trans Geol Soc S Afr 88: 267-280

Smith IE (1972) High-potassium intrusives from southeastern Papua. Contrib Mineral Petrol 34: 167-176

Smith RE, Smith SE (1976) Comments on the use of Ti, Zr, Y, Sr, K, P and Nb in classification of basaltic magmas. Earth Planet Sci Lett 32: 114-120

Smolonogov S, Marshall B (1993) A genetic model for the Woodcutters Pb-Zn-Ag orebodies, Northern Territory, Australia. Ore Geol Rev 8: 65-88

Solomon M, Groves DI (1994) The geology and origin of Australia's mineral deposits. Clarendon Press, New York, 951 pp (Oxford Mono Geol Geophys 24)

Sombroek H (1985) Igneous petrology of the Porgera intrusive "complex". BSc Honours Thesis, University of Sydney

Sørensen H (1974) Origin of the alkaline rocks; a summary and retrospect. In: The alkaline rocks, petrogenesis, pp 535-539. John Wiley and Sons, London

Spear J.A (1984) Micas in igneous rocks. In: Bailey SW (ed) Micas. American Mineralogical Society, Michigan, pp 299-349 (Rev Mineral 13)

Spies O, Lensch G, Mihm A (1984) Petrology and geochemistry of the post-ophiolitic Tertiary volcanics between Sabzevar and Quchan, NE-Iran. Neues Jb Geol Palaeont Abh, 16: 389-408

Spooner ETC (1993) Magmatic sulphide/volatile interaction as a mechanism for producing chalcophile element enriched, Archaean Au-quartz hydrothermal ore fluids. Ore Geol Rev 7: 359-379

Stanton RL (1994) Ore elements in arc lavas. Clarendon Press, New York, 391 pp (Oxford Mono Geol Geophys 29)

Stern RA, Syme EC, Bailes AH, Lucas SB (1995) Paleoproterozoic (1.90-1.86 Ga) arc volcanism in the Flin Flon Belt, Trans-Hudson Orogen, Canada. Contrib Mineral Petrol 119: 117-141

Stern RJ (1979) On the origin of andesite in the northern Mariana island arc: implications from Agrigan. Contrib Mineral Petrol 68: 207-219

Stern RJ, Bloomer SH, Lin PG, Ito E, Morris J (1988) Shoshonitic magmas in nascent arcs: new evidence from submarine volcanoes in the northern Marianas. Geology 16: 426-430

Stoffregen R (1987) Genesis of acid-sulfate alteration and Au-Cu-Ag mineralization at Summitville, Colorado. Econ Geol 82: 1575-1591

Stollery G, Borcsik M, Holland HD (1971) Chlorine in intrusives: a possible prospecting tool. Econ Geol 66: 361-367

Stolz AJ, Varne R, Wheller GE, Foden JD, Abbott MJ (1988) The geochemistry and petrogenesis of K-rich alkaline volcanics from the Batu Tara volcano, eastern Sunda arc. Contrib Mineral Petrol 98: 374-389

Strecker MR, Cerveny P, Bloom AL, Malizia D (1989) Late Cenozoic tectonism and landscape development in the foreland of the Andes: northern Sierras Pampeanas (26°-28° S), Argentina. Tectonics 8: 517-534

Stuart-Smith PG, Needham RS, Wallace DA, Roarty MJ (1986) McKinley River, Northern Territory: 1:100000 geological map commentary. Department Mines and Energy Northern Territory, Darwin

Stults A (1985) Geology of the Bajo de la Alumbrera porphyry copper-gold prospect, Catamarca Province, Argentina. MSc Thesis, The University of Arizona, Tucson

Sun SS, McDonough WF (1989) Chemical and isotopic systematics of oceanic basalts: implications for mantle composition and processes. In: Saunders AD, Norry MJ (eds), Magmatism in the ocean basins. Geological Society, London, pp 313-345 (Geol Soc Spec Publ 42)

Sun SS, Wallace DA, Hoatson DM, Glikson AY, Keays RR (1991) Use of geochemistry as a guide to platinum group element potential of mafic-ultramafic rocks: examples from the western Pilbara and Halls Creek Mobile Zone, Western Australia. Precambr Res 50: 1-35

Syme EC, Bailes AH (1993) Stratigraphic and tectonic setting of Early Proterozoic volcanogenic massive sulphide deposits, Flin Flon, Manitoba. Econ Geol 88: 566-589

Tatsumi Y, Koyaguchi T (1989) An absarokite from a phlogopite lherzolite source. Contrib Mineral Petrol 102: 34-40

Taylor SR, Capp AC, Graham AL, Blake DH (1969a) Trace element abundances in andesites II.: Saipan, Bougainville and Fiji. Contrib Mineral Petrol 23: 1-26

Taylor WR, Rock NMS, Groves DI, Perring CS, Golding SD (1994) Geochemistry of Archean shoshonitic lamprophyres from the Yilgarn Block, Western Australia: Au abundance and association with gold mineralization. Appl Geochem 9: 197-222

Thompson RN (1977) Primary basalts and magma genesis; III. Alban Hills, Roman comagmatic province, Central Italy. Contrib Mineral Petrol 60: 91-108

Thompson RN (1982) Magmatism of the British Tertiary Volcanic Province. Scott J Geol 18: 50-107

Thompson RN (1985) Asthenospheric source of Ugandan ultrapotassic magma? J Geol 93: 603-608

Thompson TB (1992) Mineral deposits of the Cripple Creek district, Colorado. Mining Engng 44 (2): 135-138

Thompson RN, Morrison MA, Hendry GL, Parry SJ (1984) An assessment of the relative roles of crust and mantle in magma genesis: an elemental approach. Phil Trans R Soc Lond A310: 549-590

Thompson JFH, Lessman J, Thompson AJB (1986) The Temora gold-silver deposit: a newly recognized style of high sulfur mineralization in the lower Paleozoic of Australia. Econ Geol 81: 732-738

Thomson BP (1969) Precambrian crystalline basement. In: Parkin LW (ed), Handbook of South Australian geology. Geological Survey of South Australia, Adelaide, pp 21-48

Thomson BP (1970) A review of the Precambrian and lower Paleozoic tectonics of South Australia. Trans R Soc S Aust 94: 193-221

Thorpe RS, Potts PJ, Francis PW (1976) Rare earth data and petrogenesis of andesite from the N-Chilean Andes. Contrib Mineral Petrol 54: 65-78

Titley SR (1975) Geological characteristics and environment of some porphyry copper occurrences in the Southwestern Pacific. Econ Geol 70: 499-514

Titley SR, ed (1982) Advances in geology of the porphyry copper deposits, Southwestern North America. University of Arizona Press, Tucson, 541 pp

Togashi S, Terashima S (1997) The behavior of gold in unaltered island arc tholeiitic rocks from Izu-Oshima, Fuji, and Osoreyama volcanic areas, Japan. Geochim Cosmochim Acta 61: 543-554

Toogood DJ, Hodgson CJ (1985) A structural investigation between Kirkland Lake and Larder Lake gold camps. Ontario Geol Surv Misc Paper 127: 200-205

Tschischow N (1989) Geologia y geoquimica de El Indio, IV Region, Coquimbo, Chile: implicancias geneticas. BSc Thesis, Universidad de Chile, Santiago

Tucker DH, Stuart DC, Hone IG, Sampath N (1980) The characteristics and interpretation of regional gravity, magnetic and radiometric surveys in the Pine Creek Geosyncline. In: Ferguson J, Goleby AB (eds) Proceedings of international uranium symposium on the Pine Creek Geosyncline, Sydney, 1979. International Atomic Energy Agency, Vienna, pp 101-140

Van Bergen MJ, Ghezzo C, Ricci CA (1983) Minette inclusions in the rhyodacitic lavas of Mt. Amiata (central Italy): mineralogical and chemical evidence of mixing between Tuscan and Roman type magmas. J Volcanol Geotherm Res 19: 1-35

Van Kooten G (1980) Mineralogy, petrology and geochemistry of an ultrapotassic basalt suite, Central Sierra Nevada, California, U.S.A. J Petrol 21: 651-684

Van Nort SD, Atwood GW, Collinson TB, Potter DR (1991) The geology and mineralization of the Grasberg porphyry copper-gold deposit, Irian Jaya, Indonesia. Mining Engng 43 (3): 300-303

Velde D (1975) Armalcolite-Ti phlogopite-diopside-analcite-bearing lamproites from Smoky Butte, Montana. Am Mineral 60: 566-573

Venturelli G, Fragipane M, Weibel M, Antiga D (1978) Trace element distribution in the Cainozoic lavas of Nevado Caropuna and Andagna Valley, Central Andes of Southern Peru. Bull Volcanol 41: 213-228

Venturelli G, Thorpe RS, Dal Piaz GV, Del Moro A, Potts PJ (1984) Petrogenesis of calc-alkaline, shoshonitic and associated ultrapotassic Oligocene volcanic rocks from the Northwestern Alps, Italy. Contrib Mineral Petrol 86: 209-220

Venturelli G, Capedri S, Barbieri M, Toscani L, Salvioli-Mariani E, Zerbi M (1991) The Jumilla lamproite revisited: a petrological oddity. Eur J Mineral 3: 123-145

Vielreicher RW, Groves DI., Ridley JR, McNaughton NJ (1994) A replacement origin for the BIF-hosted gold deposit at Mt. Morgans, Yilgarn Block, Western Australia. Ore Geol Rev 9: 325-347

Vila T, Sillitoe RH (1991) Gold-rich porphyry systems in the Maricunga Belt, Northern Chile. Econ Geol 86: 1238-1260

Villemant B, Michaud V, Metrich N (1993) Wall rock-magma interactions in Etna, Italy, studied by U-Th disequilibrium and rare earth element systematics. Geochim Cosmochim Acta 57: 1169-1180

Vollmer R, Norry MJ (1983) Possible origin of K-rich volcanic rocks from Virunga, East Africa, by metasomatism of continental crustal material: Pb, Nd and Sr isotopic evidence. Earth Planet Sci Lett 64: 374-385

Von Pichler H (1970) Italienische Vulkangebiete I. Somma-Vesuv, Latium, Toscana. Gebrüder Bornträger, Berlin

Vukadinovic D, Edgar AD (1993) Phase relations in the phlogopite-apatite system at 20 kbar — implications for the role of fluorine in mantle melting. Contrib Mineral Petrol 114: 247-254

Wagner C, Velde D (1986) Lamproites in North Vietnam? A re-examination of cocites. J Geol 94: 770-776

Waite KA, Keith JD, Christiansen EH, Whitney JA, Hattori K, Tingey DG, Hook CJ (1997) Petrogenesis of the volcanic and intrusive rocks associated with the Bingham Canyon porphyry Cu-Au-Mo deposit, Utah. In: John DA, Ballantyne GH (eds) Geology and ore deposits of the Oquirrh and Wasatch Mountains, Utah. Econ Geol Guidebook Series 29: 69-90

Walker RT, Walker WJ (1956) The origin and nature of ore deposits. Walker Associates, Colorado Springs, unpaginated

Wall VJ (1990) Fluids and metamorphism. PhD Thesis, Monash University, Melbourne

Wallace DA, Johnson RW, Chappell BW, Arculus RJ, Perfit MR, Crick IH (1983) Cainozoic volcanism of the Tabar, Lihir, Tanga, and Feni islands, Papua New Guinea: geology, whole-rock analyses, and rock-forming mineral compositions. Bur Mineral Resources, Austr, BMR Rep 243

Walthier TN, Sirvas E, Araneda R (1985) The El Indio gold, silver, copper deposit. Engineer Min J 186(10): 38-42

Warnaars FW, Smith WH, Bray RE, Lanier G, Shafiqullah M (1978) Geochronology of igneous intrusions and porphyry copper mineralization at Bingham, Utah. Econ Geol 73: 1242-1249

Weaver BL, Wood DA, Tarney J, Joron JL (1987) Geochemistry of ocean island basalts from the South Atlantic: Ascension, Bouvet, St. Helena, Gough and Tristan da Cunha. In: Fitton JG, Upton BGJ (eds) Alkaline igneous rocks. Geological Society, London, pp 253-267 (Geol Soc Spec Publ 30)

Webster JD (1992) Water solubility and chlorine partitioning in Cl-rich granitic systems: effects of melt composition at 2 kbar and 800°C. Geochim Cosmochim Acta 56: 679-687

Webster JD (1997) Chloride solubility in felsic melts and the role of chloride in magmatic degassing. J Petrol 38: 1793-1807

Webster JD, Holloway JR (1988) Experimental constraints on the partitioning of Cl between topaz rhyolite melt and H_2O and H_2O+CO_2 fluids: new implications for granitic differentiation and ore deposition. Geochim Cosmochim Acta 52: 2091-2105

Webster JD, Holloway JR (1990) Partitioning of F and Cl between magmatic hydrothermal fluids and highly evolved granitic magmas. In: Stein HJ, Hannah JL (eds), Ore-bearing granite systems; petrogenesis and mineralizing processes. Geological Society of America, pp 21-33 (Geol Soc Am Spec Paper 246)

Welsh TC (1975) Cadia copper-gold deposits. In: Knight CL (ed) Economic geology of Australia and Papua New Guinea. Australasian Institute of Mining and Metallurgy, Parkville, pp 711-716 (Australas Inst Min Metall Monogr 5)

Wheller GE (1986) Petrogenetic studies of basalt-andesite-dacite volcanism at Batur volcano, Bali, and the causes of K-variation in Sunda-Banda-Arc basalts. PhD Thesis, University of Tasmania, Hobart

Wheller GE, Varne R, Foden JD, Abbott M (1986) Geochemistry of Quaternary volcanism in the Sunda-Banda arc, Indonesia, and three-component genesis of island arc basaltic magmas. J Volcanol Geotherm Res 32: 137-160

White NC, Leake MJ, McCaughey SN, Parris BW (1995) Epithermal gold deposits of the Southwest Pacific. J Geochem Expl 54: 87-136

White WH, Bookstrom AA, Kamilli RJ, Ganster MW, Smith RP, Ranta DE, Steininger RC (1981) Character and origin of Climax-type molybdenum deposits. Econ Geol 75th Anniv Vol: 270-316

Whitford DJ (1975) Geochemistry and petrology of volcanic rocks from the Sunda arc, Indonesia. PhD Thesis, The Australian National University, Canberra

Whitford DJ, Jezek PA (1979) Origin of Late-Cenozoic lavas from the Banda arc, Indonesia: trace element and Sr isotope evidence. Contrib Mineral Petrol 68: 141-150

Whitford DJ, Nicholls IA, Taylor SR (1979) Spatial variations in the geochemistry of Quaternary lavas across the Sunda Arc in Java and Bali. Contrib Mineral Petrol 70: 341-356

Wilson F (1981) The conquest of Copper Mountain. Atheneum, New York, 244 pp

Wilson JC (1978) Ore fluid-magma relationships in a vesicular quartz latite porphyry dyke at Bingham, Utah. Econ Geol 73: 1287-1307

Wilson M (1989) Igneous petrogenesis; a global tectonic approach. Unwin Hyman, London, 466 pp

Wolfe RC (1999) The geology and petrogenesis of the alkaline Didipio intrusive complex, North Luzon, Philippines. Mineral Petrol (in press)

Wolfe RC, Cooke DC, Joyce P (1998) The Dinkidi Au-Cu porphyry - an alkaline porphyry deposit from North Luzon, Philippines. Geol Soc Austr Abst 49: 131

Wood DA, Joron JL, Treuil M (1979) A re-appraisal of the use of trace elements to classify and discriminate between magma series erupted in different tectonic settings. Earth Planet Sci Lett 45: 326-336

Woodhead JD (1989) Geochemistry of the Mariana arc (western Pacific): source composition and processes. Chem Geol 76: 1-24

Woodhead J, Eggins S, Gamble J (1993) High field strength and transition element systematics in island arc and back-arc basin basalts: evidence for multi-phase melt extraction and a depleted mantle wedge. Earth Planet Sci Lett 114: 491-504

Wörner G, Harmon RS, Davidson J, Moorbath S, Turner DL, McMillan N, Nye C, Lopez-Escobar L, Moreno H (1988) The Nevados de Payachata volcanic region (18° S/69° W, N. Chile). Bull Volcanol 50: 287-303

Wyborn D (1988) Ordovician magmatism, Au mineralization and an integrated tectonic model for the Ordovician and Silurian history of the Lachlan Fold Belt, NSW. BMR Res. Nletter 8: 13-14

Wyborn D (1992) The tectonic significance of Ordovician magmatism in the eastern Lachlan Fold Belt. Tectonophysics 214: 177-192

Wyborn D (1994) Mantle magmatism and large gold-copper deposits. In: Ore deposit studies and exploration models: porphyry copper-gold deposits, short course manual. CODES, University of Tasmania, Hobart, pp 5.1-5.4

Wyman D (1990) Archean shoshonitic lamprophyres of the Superior Province, Canada, petrogenesis, geodynamic setting, and implications for lode gold deposits. PhD Thesis, University of Saskatchewan, Saskatoon

Wyman D, Kerrich R (1988) Alkaline magmatism, major structures and gold deposits: implications for greenstone belt gold metallogeny. Econ Geol 83: 454-461

Wyman D, Kerrich R (1989a) Archean shoshonitic lamprophyres associated with Superior Province gold deposits: distribution, tectonic setting, noble metal abundances and significance for gold mineralization. In: Keays RR, Ramsay WRH, Groves DI (eds), The geology of gold deposits: the perspective in 1988. The Economic Geology Publishing Company, El Paso, pp 651-667 (Econ Geol Monogr 6)

Wyman D, Kerrich R (1989b) Archean lamprophyre dykes of the Superior Province, Canada: distribution, petrology and geochemical characteristics. J Geophys Res 94: 4667-4696

Yang K, Bodnar RJ (1994) Magmatic-hydrothermal evolution in the "bottoms" of porphyry copper systems: evidence from silicate melt and aqueous fluid inclusions in granitoid intrusions in the Gyeongsang Basin, South Korea. Int Geol Rev 36: 608-628

Yeats CJ, McNaughton NJ, Ruettger D, Groves DI (1999) Evidence for two Archean lode-gold mineralizing events in the Eastern Goldfields of the Yilgarn Craton, Western Australia: a SHRIMP U-Pb study of felsic porphyries. Econ Geol (in press)

Zhang M, Suddaby P, Thompson RN, Thirlwall MF, Menzies MA (1995) Potassic volcanic rocks in NE China: geochemical constraints on mantle source and magma genesis. J Petrol 36: 1275-1303

Zhou JC, Zhou JP, Liu J, Zhao TP, Chen W (1996) Copper (gold) and non-metal deposits hosted in Mesozoic shoshonite and K-rich calc-alkaline series from Lishui in the Lower Yangtze region, China. J Geochem Expl 57: 273-283

Subject Index

structural element *continued*
  thrust 115, 122
  trough 106
structural setting 186-198, 200-209
subduction 2, 7, 11, 18, 22, 45, 46, 47, 55,
    56, 57, 69, 71, 95, 99, 101, 107, 122,
    127, 140, 141, 161, 165, 182, 183,
    185
  angle 13, 18, 19, 39, 115
  arc/zone 69, 115, 155, 185
  depth 19
  rate 39
  ridge, *see* ridge
  trench, *see* trench
subvolcanic, *see* hypabyssal
sulphur
  saturation 69, 70, 170
  undersaturation 68, 69, 70

tectonic element
  basin 95, 101
  nappe 47
  platform 145
  rift 5, 8, 11, 63, 145
  setting 13-19
    continental 13
    oceanic 13
  slab 2, 22, 23, 37, 115, 140, 183
  suture 47, 182
  underthrust 161
  uplift 115, 117, 137
tectonic regime 18
  accretion 161
  compression 18, 47, 115, 157
  extension 18, 47, 107, 135
  rifting 5, 18, 19, 40, 57
  transpression 47, 135, 161, 182
terrain
  basement 47, 103, 115, 117, 122
  gneiss 152, 155
  granitoid-greenstone 154, 155, 160, 161
  greenstone belt 68, 155, 157, 160, 161
  igneous complex/province, *see* igneous
    rock
  sedimentary sequence, *see* sedimentary
    rock
  supracrustal 155
  volcanic sequence, *see* volcanic rock
  volcanosedimentary sequence, *see*
    volcanosedimentary rock
terrane 106, 153, 155, 161
  ancient 2, 11

terrane *continued*
  Kosciusko, New South Wales, Australia,
    see geographic/geological names
  Yilgarn Block, Western Australia, *see* geo-
    graphic/geological names
texture
  aphanitic 128
  aplitic 128
  arborescent 157
  banded 125
  battlement structure 7, 73, 75, 157
  colloform 125
  crenulated 132
  cryptocrystalline 125
  drusy 125
  equigranular 12, 109, 121, 135, 188, 194,
    195, 197, 198, 201, 206
  flow 73, 75
  foliated 132
  globular structure 7, 75
  glomeroporphyritic 64
  holocrystalline 111
  layered 202
  massive 95, 125, 132, 191, 192, 193
  miarolitic cavities 182
  mosaic 107
  ocellar 157
  ophitic 139
  phaneritic 200
  porphyritic 6, 7, 9, 12, 45, 59, 61, 64, 73,
    97, 100, 105, 107, 117, 128, 130, 135,
    139, 181, 186, 187, 189, 190, 191,
    192, 194, 195, 196, 197, 198, 199,
    200, 201, 202, 203, 204, 205, 206,
    207, 208, 209
  vughy 105
  xenolithic 128
  zoned 53, 73, 75, 79
TNT anomaly, *see* spidergram
trench 7, 18, 31, 39, 57, 58, 59, 61
  Cotobato Trench 102
  Luzon Trench 101
  Manila Trench 101, 102
  Negros Trench 102
  New Hebrides Trench, *see* Vanuatu Trench
  Philippine Trench 102
  Tonga-Kermadec Trench 98
  Vanuatu Trench 99
  Vitiaz Trench 99
trough
  East Luzon Trough 102

volatile 6, 7, 8, 37, 39, 40, 69, 97, 103, 104,
        105, 107, 117, 137, 139, 141, 153,
        157, 166, 167, 170, 182, 183
    chlorine, *see* halogen, chlorine
    carbon dioxide 8, 34, 40, 64, 68, 151, 157,
        161
    fluorine, *see* halogen, fluorine
    hydrogen 69
    hydrous mineral 117, 171
    loss on ignition (LOI) XII, 27, 84, 93
    water 8, 34, 64, 69, 169, 171, 182, 183
volcanic, *see* depth of formation
volcano 45, 56, 57, 63, 64
    Batu Tara, *see* geographic/geological
        names
    caldera 64, 95, 97, 99, 101, 104, 105, 107,
        141, 182, 193, 194, 195, 197, 205

volcano *continued*
    dome 122
    Karisimbi, *see* geographic/geological
        names
    Nyiragongo, *see* geographic/geological
        names
    shield 59, 64, 99
    subaerial 60, 95
    submarine 57, 60
    strato 95, 117, 122

within-plate setting 7, 9, 10, 11, 13, 14, 16,
        17, 18, 19, 22, 25, 36, 40, 46, 63, 66,
        68, 71, 81, 89, 90, 91, 139, 140, 141,
        150, 152, 153, 166, 173, 184, 191,
        206

Printing: Mercedes-Druck, Berlin
Binding: Buchbinderei Lüderitz & Bauer, Berlin